T0133949

OPEC and the World's Energy Future

OPEC and the World's Energy Future

Its Legacy and Promise

Mohammed A. Alsahlawi

CRC Press
Taylor & Francis Group
Boca Raton London New York

CRC Press is an imprint of the
Taylor & Francis Group, an **informa** business

First edition published 2021
by CRC Press
6000 Broken Sound Parkway NW, Suite 300, Boca Raton, FL 33487-2742

and by CRC Press
2 Park Square, Milton Park, Abingdon, Oxon, OX14 4RN

Library of Congress Cataloging-in-Publication Data

Names: Alsahlawi, Mohammed A., 1954- author.
Title: OPEC and the world's energy future : a 60-year legacy and promise /
author: Mohammed A. Alsahlawi.
Description: First edition. | Boca Raton, FL : CRC Press, 2021. | Includes
bibliographical references and index. | Summary: "This work offers a
complete account of OPEC's past, present, and possible future in
relation to economic, political, and technological changes. It focuses
on the impacts of recent international political and economic
developments and analyzes factors affecting OPEC as well as the world
oil market. The book describes the continued importance of oil and gas
as major sources of energy throughout the world. It examines OPEC's
history and merits, highlights differences among OPEC members, and
discusses OPEC's relations with the world. It spends considerable time
on concentrating on the impact of new technologies and how they may
challenge and change the organization in the near and long term"--
Provided by publisher.
Identifiers: LCCN 2020055199 (print) | LCCN 2020055200 (ebook) | ISBN
9780367349783 (hardback) | ISBN 9780367342142 (paperback) | ISBN
9780429328213 (ebook)
Subjects: LCSH: Organization of Petroleum Exporting Countries--History. |
Petroleum industry and trade. | International agencies. | International
trade. | Cartels. | Energy policy.
Classification: LCC HD9560.1.O66 A697 2021 (print) | LCC HD9560.1.O66
(ebook) | DDC 382/.42282--dc23
LC record available at https://lccn.loc.gov/2020055199
LC ebook record available at https://lccn.loc.gov/2020055200

ISBN: 978-0-367-34978-3 (hbk)
ISBN: 978-0-429-32821-3 (ebk)

Typeset in Times
by Deanta Global Publishing Services, Chennai, India

Contents

Preface

Since the early years of the last century, international oil companies (the Seven Sisters) have been dominating the oil industry and controlling oil prices. As much as oil demand has increased over the years, new oil discoveries have been developed in different parts of the world, mainly in the Middle East and Latin America. At the same time oil companies have successively reduced the oil price due to competition and mistrust among them and due to increased oil supplies from Russia.

With the increase of nationalism and anti-imperialism as features of independence movements in the colonies, two oil-producing countries – Saudi Arabia and Venezuela – coordinated their efforts against oil companies in order to put a limit on oil price decline and safeguard their oil revenues. The struggles of oil producers in the face of international oil companies set the stage for the creation of the Organization of the Petroleum Exporting Countries (OPEC).

In September 1960, OPEC was established with the purpose of coordinating petroleum policies between its members. In September 2020, OPEC celebrates its sixty-year anniversary. Over the first decade of its foundation, OPEC was met by ignorance and was called a cartel with no significant achievements. During the 1970s and 1980s, OPEC was taking the leading role in setting oil prices and controlling the oil market.

Over the years, OPEC was facing ever changing political challenges and market structural changes such as emerging oil futures. Yet, OPEC was – and still is – highly affected by political and economic events and developments, which have been the determinants of its future and its role as a key player in the world oil market.

This book reviews the historical causes and reasons for forming OPEC and discusses the critical paths of its evolution, given the surrounding geopolitics. The aim and scope of this book is to identify the forces that shape the world's energy future and where OPEC could stand. The main question is how OPEC affects energy's future or vice versa and what strategies might be adopted in facing the new energy changes and challenges, which certainly raise the question of its merit and existence. The energy and technology interaction will be an essential element in drawing the future of OPEC as well as the world's energy future.

The book provides a recent review of regional geopolitical developments and highlights the differences and relations between OPEC members and between OPEC and outsiders. The importance of the book is stemmed from the author's personal views as an insider of being a director of the OPEC Information Department and the OPEC News Agency (OPECNA) during the 1990s. The author was responsible for organizing media coverage of OPEC conferences and monitoring in-out OPEC news. Several side issues have received high international and domestic media coverage. As such, OPEC was so keen in covering those issues because of its importance and relevance.

The most recent issue is the global pandemic coronavirus which erupted in early 2020 and given the name COVID-19. The world was put under lockdown where most social and economic activities have been shut down and the world economy went into a deep recession. The newly implemented rules and regulations have altered the world into a new world order. The negative economic impacts caused a sluggish oil demand and created a unique challenge to OPEC in dealing with oil over supply and controlling the oil price decline. Until now, there is no serious remedy for the coronavirus, and even with proposed vaccines, there are debates as to its effectiveness and safety. The most definite thing is the lasting effect of COVID-19 on human health and the long-term political and economic impact. The issue has been turned into a trade-off between a COVID-19 lockdown and a prosperous economy.

The OPEC insider experience of the author has put him in face of important managerial and diplomatic situations. These have ranged from dealing with day-to-day business to encountering very delicate issues. The day-to-day business was typically dealing and working with different OPEC officials from different member countries who work as officers at the OPEC secretariat and supposed to represent their countries and advocate their political and economic interests.

Other important minute-by-minute business matters were responding to the media and press correspondents' questions and enquiries or news items pertaining to OPEC. This is aside from the regular activities of organizing press conferences during OPEC ministerial meetings and other professional meetings or seminars and workshops related to oil and energy matters. Other essential activities, which the author was responsible for, are supervising, writing, and editing of OPEC deferent publications and newsletters. The most critical issue related to OPEC publications is using the notion of the Gulf rather than the Arabian or Persian Gulf to avoid any conflict or sensitivity between Arab countries and Iran.

The most important and sensitive issue that the author had to deal with while working with OPEC in the 1990s was the environmental problem. This problem has developed a great deal of debate between oil producers as represented by OPEC and the environmentalists especially industrialized oil-consuming countries and other organizations such as Greenpeace. For example, in 1992, OPEC organized an international seminar on the importance and effect of environmental measures on OPEC oil which was held at the OPEC Secretariat.

For this event, the author was responsible for organizing and other supervizing management activities. In the early morning of the first day of the seminar, when the author reached the OPEC premises, the author was surprised and astonished by seeing the whole building covered by a big poster showing the Greenpeace themes and logos and several of its activists were hanging by ropes around the building, protesting against OPEC and fossil fuels.

Before starting the seminar, the OPEC secretary general at that time Dr. Subrotto from Indonesia asked the author to start negotiations with the protestors. After negotiations, the author succeeded to induce them to participate in the seminars and start the opening session by a short speech by their leader for about ten minutes to speak about the importance of preserving the environment.

Such incidents will add to the positive features of the book and to its credibility and depth. Furthermore, the book has the merit to show that diplomacy and proper care of handling uprisings will make everything possible. The recent and most challenging world and regional geopolitical event is the normalization of diplomatic relations between Israel and Arab countries. This dramatic move for sure will change the political and economic landscapes of the region and will create new challenges for OPEC and its influence on the world oil market.

Mohammed A. Alsahlawi

Acknowledgments

The author would like to thank OPEC at ministerial and administrative levels who were an inspiration for the work.

The author would like to thank Dr. Jennifer Considine for her encouragement throughout writing the book and her constructive suggestions.

The author would also like to thank Mr. Peter Adam for his excellent ideas on the title of the book. Thanks also go to Mr. Junaid Akhtar for his typing assistance.

Special appreciation goes to my family for the support and patience during my work abroad at OPEC.

Author

Mohammed A. Alsahlawi is Professor of Economics and Energy Economics, KFUPM Business School, (formerly College of Industrial Management), King Fahd University of Petroleum and Minerals (KFUPM), Saudi Arabia, and previous Dean of the College of Industrial Management. He holds a Ph.D. in economics from the University of Wisconsin-Milwaukee, US (1985), B.S. in chemical engineering (1978), and an MBA (1980) from KFUPM, Saudi Arabia. In 1985 he was the Director of the Economic and Industrial Research Division, Research Institute, KFUPM, Saudi Arabia. Alsahlawi was the Director of the Organization of the Petroleum Exporting Countries (OPEC) Information Department and OPEC News Agency from 1991 to 1995 and was a member of the first advisory board of the Saudi Arabian Supreme Economic Council from 1999 to 2002. He established and was the director of the Human Resources Development Fund (HRDF) from 2001 to 2006. He also served as consultant to major industrial and business firms and governmental organizations in Saudi Arabia from 1985 until present. Alsahlawi serves on several editorial boards of international journals in energy economics and business economics, and his publications have appeared in several energy economics journals while his specialized comments and writings have appeared in different news and trading magazines.

PART

I Introduction

1.1 PRE-OPEC

Since the mid-nineteenth century, oil wells have been drilled on a commercial basis in the United States. In fact, Lord Cochrane, the Earl of Dundonald's great grandfather, had the first commercial patent for the use of oil and the birth of the world oil industry (The Westminster Review, vol. 21–22).

Europe and Russia have followed track, and with the increased importance of oil and more uses of oil having been introduced, oil has been discovered in new parts of the world. Refined oil product such as kerosene and middle distillates have been developed and put in use. Until the first two decades of the twentieth century, the United States was the leading country in oil production, refining, and consumption. The Middle East, on the other hand, shared less than 5 percent of the world's production in 1960. This share has increased to around 33.5 percent in 2018 as shown in Table 1.1 (Alsahlawi, 2014).

Spontaneously, the United States had the power of controlling oil prices in the world. However, the pricing system was not in the hand of the government, but under the control of major oil companies, which played as a tacit cartel in setting oil prices while the rest of world's oil companies had to follow.

Before World War I, the world oil market was dominated by four major international oil companies: Shell, Standard Oil, Nobel, and Rothschild.

The latter two companies were in Russia and were liquidated as private companies by the 1917 Russian Revolution. Another major company founded by the British government, was the Anglo–Persian Oil Company (now British Petroleum). In the 1920s, the oil market was essentially controlled by these three companies. In the 1930s, new major oil companies developed as offshoots of the old standard oil company. They were Gulf, Texaco, Standard of California, Sohio, and Mobil.

With these new entrants, the degree of competition in the world oil market increased, but only to a certain extent. In the 1940s and 1950s, the Seven Sisters (Gulf, Texaco, Standard of California, Sohio, Mobil, British Petroleum, Shell) had balanced the supply and demand mainly by market-sharing and joint producing agreement. To some extent these agreements distorted world market competition, resulting in an oligopoly market structure characterized by substantial differences between production cost and market price.

The positive differential of oil prices from production costs allowed for vertical integration and controlling the market all the way from exploration to marketing. The share of the major oil companies in world oil production refining and marketing was about 60 percent.

This concentration ratio, which indicates the degree of monopoly in the world oil market, has decline dramatically, especially in the production sector.

TABLE 1.1

Share of World Crude Oil Production by Region (mbd), 1960 to 2018

Region	1960	1990	2018	2018 Share of Total
North America	9.20	13.85	22.58	23.86%
Latin America	2.90	4.51	6.69	6.9%
Western Europe	0.30	3.70	3.73	3.7%
Eastern Europe	3.20	12.4	14.48	15.3%
Middle East	5.30	17.54	31.76	33.5%
Africa	0.28	6.72	8.19	8.6%
Asia and Pacific	0.60	6.73	7.63	8.1%
Total	**21.78**	**65.46**	**94.72**	**100%**

Source: BP Statistical Review of World Energy, London, 2018. With permission.

This is due to the increased participation of oil-producing countries in production and to the evolution of the national oil companies. The market power of the majors has reduced but they still control the refining and marketing sectors and participate in production from technical side. On some occasions, the federal government agents. Mainly in the US, the Federal Treadle Commission (FTC) interferes in setting oil pricing if a monopoly power exists. During both world wars, the importance of oil became undoubtable and undeniable.

Subsequently, competition arose among international oil companies and between the main colonial powers. It became severe to the extent that oil became the source of conflicts. Since WWII, new political and economic systems have emerged, and new uses of oil have been developed to fulfil new needs and new economic development requirements. Regardless of world oil in the newly developed regions and countries which altered oil production and consumption, the patterns of power for controlling the oil price stayed in the hands of international oil companies – mainly the Americans – despite the survival of long-standing European oil companies, such as British Petroleum and Shell.

These companies maintained a good grip on the oil price because of the vertical advantages over different stages of the oil industry – from integration, exploration, and development to downstream sectors of refining, pricing, and distribution, as in Figure 1.1.

The competition between these companies appeared clear in securing oil and gas concessions in different countries all over the world.

There were – and still are – great prospects for oil and gas in the Middle East and South America. As a matter of fact, countries such as Saudi Arabia, Kuwait, and Iraq, have voluntarily offered a generous concession to international oil companies. The American companies are concentrated in Saudi Arabia, while the European companies were historically present in Iraq, Kuwait, and Iran. With the increasing importance of oil and increasing competition between oil companies over the countries with oil potential.

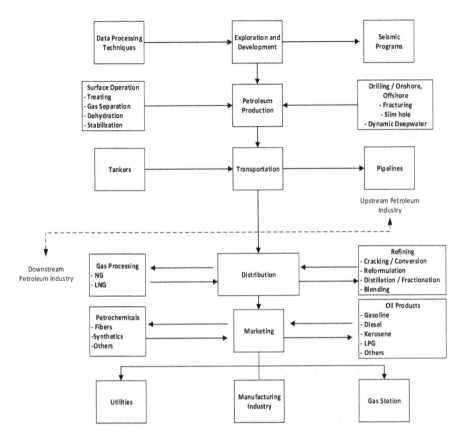

FIGURE 1.1 Petroleum industry stages. Source: Alsahlawi, 2014.

The prices declined to levels that did not match the rapid rise in world oil demand. This put low price economic and social pressures on oil-producing countries in the Middle East and South America. Such pressures added to the nationalism movements that widely spread in different regions and became phenomena of post-colonialism era.

With an increase in independence movements and nationalism during the 1950s, international oil companies became a target in a way to retaliate from imperialism. This set justified grounds to question the low price and unfair concession deals.

With successive cuts in oil price by the oil companies, Venezuela and Saudi Arabia coordinated their efforts to put some control on oil prices not to go further down in a way to safeguard their revenues. In parallel, the Arab oil-producing countries decided to meet at the Arab Petroleum Congress in Cairo in April 1959.

The resolution came out of the Arab Petroleum Congress meeting to call for more coordination and not to take unilateral decisions on oil prices cut by international oil companies. Venezuela attended that meeting as an observer and offered a proposal to form an oil consultative committee.

This was the seed for OPEC's creation, after the successive price cut by oil companies led by Exxon. Given the rise of oil production from Russia which continued in the 1960s, where Soviet president Khrushchev was using oil as an economic weapon in the cold war (Learsy, 2007).

1.2 OPEC EVOLUTION

The mistrust among oil companies and among oil-producing countries led to the formation of OPEC. Saudi Arabia and Venezuela set the coordination between oil producers in the Middle East and Latin America in the face of international oil companies and they succeeded in convincing Iran, Iraq, and Kuwait to join the effort. On 14 September 1960, the five countries met in Baghdad and announced the formation of OPEC, see Figure 1.2.

The main purpose of OPEC as declared in that meeting was to coordinate petroleum policies between member countries and to safeguard their interests.

FIGURE 1.2   1st OPEC conference, September 10–14, 1960, Baghdad, Iraq. Source: www .opec.org

Qatar joined in 1961 and was followed by Indonesia and Libya in 1962. The United Arab Emirates, Algeria, Nigeria, Ecuador, and Gabon joined in 1967, 1969, 1971, 1973, and 1975 respectively. The number totaled 13 in 1995. However, over time some countries suspended or terminated their memberships By 2019, the members totaled 14, see Figure 1.3.

OPEC chose its headquarters in Geneva in order to be perceived as an international neutral body with equal votes and financial contributors from the members. In June 1965, OPEC moved to Vienna by invitation from the government of Austria with better diplomatic status and privileges. Figure 1.4 shows the signing of the headquarter agreement between the Government of Austria and OPEC.

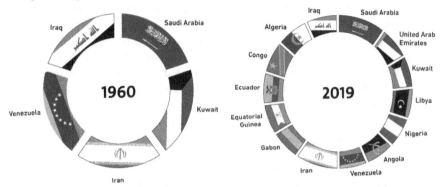

FIGURE 1.3   OPEC member countries, 1960, 1920.

June 24, 1965, Vienna, Austria: Headquarters Agreement between the Government of Austria and OPEC. Dr Bruno Kreisky, Austrian Foreign Minister, and Dr. Ashraf Lutfi, OPEC Secretary General.

FIGURE 1.4   The headquarters agreement. Source: http://www.opec.org

OPEC's aim was to counteract the monopoly power of oil companies. In that effort, western media and oil companies considered OPEC as a cartel. However, nobody took OPEC seriously, especially oil companies who persistently continued in lowering oil prices.

Over the years, OPEC membership has increased but there was no clear or solid achievement toward stabilizing oil prices. In the 1960s, the oil market was still structured as a monopoly dominated by oil companies as before. However, OPEC tried to improve the economic wealth of its members through negotiating for higher prices, and more involvement in the oil pricing mechanism.

This was not attained for many reasons; most importantly, the oil producers had poor know-how of the petroleum industry and business, which were in the hands of oil companies.

The first effective step taken by OPEC towards increasing the oil price was in 1971 as a result of an oil demand rise and after signing the Tehran agreement in February 1971. However, oil prices were still posted by oil companies until OPEC took the initiative to apply OPEC official prices and used the Saudi Arabian light crude oil as an official price after October 1973.

This OPEC official price was a reaction to the changes in market conditions and to spot oil prices until OPEC adopted market-based pricing (the oil pricing benchmark or OPEC's reference basket price in the late 1980s).

Over the first decade of OPEC's foundation, it received the notion of a cartel. This idea has widely spread and is quoted in both press and academia.

It is hard to find any basic economic textbook that does not refer to OPEC as a classic example of a cartel. This is strongly endorsed by oil companies, which are OPEC's primary enemy. In spite of tacit agreement among western governments for OPEC's creation, they allied with oil companies in most bilateral debates. During the decade of the 1970s, OPEC appeared on the economic and political arenas, as its members started to take some control over oil prices. This was enhanced by their initiatives to stabilize oil prices during the 1973 and 1979 price shocks. The price crash of 1986 pushed OPEC to the edge. It had to deal with the oil glut collectively and individually.

OPEC gave up its market share to non-OPEC oil producers; Saudi Arabia as the major oil producer and the biggest producer within OPEC had to play the swing producer role on account of its market share and its production quota.

This forced OPEC to change its production and prices systems, where production ceilings and a reference price were adopted. Yet, the market uncertainty continued in the 1990s and beyond.

This market instability was aggravated by the emergence of environmental issues and some macroeconomic problems such as the South-East Asian financial crisis of the mid-1990s. Climate change received high attention at international level, where the United Nations sponsored the Earth Summit of 1992 and OPEC became one of the key players in order to promote its interests in protecting the environment and advocating better use of fossil fuels. The economic and political issues of the 2000s put high pressure on OPEC to modify its strategies in a way to accept the charges in market forces and the new energy mix which skewed towards renewable energy.

Since the United Nations Conference on Environment and Development (UNCED) which was held in Rio de Janeiro in June 1992 and known as the Rio Earth Summit, the environment became an international concern. As a result of the conference, all nations stressed the idea that the world is resting on three pillars: Energy, Environment, and Economic Development. The main official documents agreed upon at the earth summit are: the convention on biological diversity and the convention on climate change or (global warming). The magic words in all outcomes of the Rio Summit is the sustainability in energy, environment, and economic development. It is a very hard task to be achieved globally. More than 150 heads of state signed the Rio conventions and endorsed the Rio Declaration and adopted Agenda 21 to ensure the sustainability principle in the 21st century. However, there were clear disputes between rich developed countries and poor less developed countries which have appeared on the real positions of those countries and on the degree of their commitments to the signed agreements. Regardless of how much the world succeeded in meeting the earth summit targets, OPEC is clearly at the heart of the matter. For the question of how to promote using fossil fuels against renewable energy, OPEC has to have a realistic argument based on convincing scientific facts. It is an intense challenge and there is a continuous struggle for OPEC to provide clean energy and preserve the environment at the same time as an engine for economic developments in the world and member countries.

PART
II OPEC's Role in World Oil Market

2.1 OPEC FINDS ITS WAY

Since the early days of OPEC, the intention of the founders was – and still is – to have an inter-governmental body representing oil producers from developing countries. There was great doubt of its survival especially with the ignorance of international oil companies and consuming industrialized countries. A very important feature of OPEC is the nature of the commodity that it deals with, which is oil: the most strategic and exhaustible commodity in existence today. The struggle of OPEC was seen in the negotiations with oil companies about pricing and production policies.

The aim was to convince oil companies and oil industry participants to have a fair oil price, which has to reflect the true value of oil. OPEC argued that oil was underpriced by the major oil companies and the return to the oil producers was only 10 percent of the value of a barrel of crude oil, while the companies and consuming countries realized almost a 90 percent return from sales and taxes.

The objectives of OPEC were – and remain – to unify the pricing and production policies among member countries while maintaining a fair price and market stability. The struggle to gain recognition and credibility was the hardest task for OPEC to deal with in the western media. This is a general mistrust held in the west toward OPEC, which stemmed from a stereotype of OPEC being represented by Arabs.

Due to persistent arguments and pressures from OPEC, and governments of oil producers, the companies accepted to discuss oil issues with the host governments. The major outcome was the agreement by the oil companies to re-calculate taxes and royalties on the basis of posted rather than realized prices. The OPEC success continued in the following years, until OPEC assumed the "sovereign right" for its members to manage their oil and petroleum resources.

Another success was the release of the Declaratory Statement of Petroleum Policy in Member Countries in the June 1968 meeting, which allowed OPEC countries to go into direct exploration and development.

As oil prices increased, non-OPEC production increased while overcoming the high cost of production. This caused OPEC to play the role of residual supplier and act as price administrator in order to protect price stability, even when OPEC's market share was eroded. In the process of OPEC trying to find its way, it continued to interact with oil companies, consuming countries, non-OPEC producers, and other international organizations such as the United Nations (UN), World Energy Council (WEC), International Energy Agency (IEA), and the Organization of Economic Cooperation and Development (OECD). In most OPEC conferences and important meetings there were representatives from these organizations.

Before instituting what was known as "OPEC and non-OPEC dialogue", OPEC organized a joint ministerial meeting between OPEC and the key Independent Petroleum Exporting Countries (IPEC) in Muscat, Oman on April 13, 1993. The IPEC/OPEC ministerial meeting was attended by 12 OPEC members countries and by 14 IPEC producers, among them the Russian Federation, Norway, Mexico, Malaysia, China, and the State of Texas. Some attended as observers from international organizations. The meeting reviewed the long-term energy and oil-market outlook in light of different policies concerning the environment and investment prospects in both the producing and consuming countries. One of the important policies pertained to OPEC and other petroleum exporting countries; there was the issue of imposing a carbon and energy tax by consuming countries. Overall, the meeting called for more cooperation among producers and consumers which eventually materialized in producer-consumer dialogue which was implemented in the 1990s and after.

The image of OPEC changed toward a positive one as a key player in the world oil market with a great ability to deal with oil and financial crises as much as needed or by its initiative. This happened during the Gulf Wars in the 1980s and early 1990s when oil supply was partially disrupted and OPEC was obliged to secure the flow of oil to the world. This tendency to react to unexpected situations **should be enhanced to plan** for long-term events that might face OPEC and determine possible ways to encounter it. In other words, OPEC needs to have a flexible and sustainable long-term strategy for dealing with oil price fluctuations and market changing conditions, at the same time as responding positively to the media.

Furthermore, OPEC orchestrated and organized several campaigns to promote its image and achievements to justify its importance and existence on different fronts, especially the environment and climate change, and eliminating world poverty. On the environmental front, OPEC as an organization or as individual members contributed financially to promote research on environmental protection and sponsored environment and climate change meetings and workshops (1992). On a fighting poverty front, OPEC established the OPEC Fund in 1975 to enhance and fund economic development aid projects and programs in less developed countries (LDCs).

Furthermore, OPEC participation in the humanities oil-for-food program in Iraq during US economic sanctions imposed on Iraq in 1995, which allowed Iraq to sell oil under United Nations' supervision and control in exchange for food, medicine, and other humanitarian needs. Such a program could be repeated and organized by OPEC to help different OPEC countries that have economic and political problems such as recently in the case of Venezuela and Libya.

2.2 INTERNAL RIVALRY

During the early years and the first decade of establishing OPEC, there was almost a general consensus and close coordination among its members. Each member was trying to preserve the main objective of attaining a reasonable high price in order to safeguard a fair oil revenue. The rivalry was mainly between OPEC as an organization and major international oil companies over the oil price change. However, internal rivalry was minimal and generally concerned with the appointment of the Secretary General, and the issue was eventually settled by annual rotation from each country. The first five secretary generals were decided to be representing the founding members. They were chosen in alphabetical order, starting with Iran in 1961 for the first three years; then the following General Secretary was appointed by the conference for one year. During the Iran–Iraq war, there was a dispute over the Secretary General's position, especially on anyone nominated from Iran or Iraq. Therefore, between 1983 and 1988, the OPEC Secretariat was led by five ministers who were asked to supervise day-to-day work in the absence of secretary generals.

Afterward the normal procedures to appoint a Secretary General every year were re-applied – except for the years when political differences appeared severe, such as the situation when Iraq invaded Kuwait, and the US imposed sanctions against Iran.

In general, whatever the disagreements between member countries over the position of the Secretary General, there were always practical compromises. Then, there was minor rivalry among the members over production levels, market shares, or the oil price levels. This is because each country has no control over the production levels – at least from an operational side. The production decision was mainly in the hands of the partners or the major international oil companies, which were responsible for operations on the basis of the concession agreements between the host countries, the member countries, and oil companies.

Over the course of the 1970s, the political scene changed with changing economic priorities for each member country. In addition to that, the number of members has increased over the years, which increased the internal rivalry among the members regarding deciding prices and output distribution.

Generally speaking, with the increasing number of member countries, the adherence to production decisions and quota allocations became difficult. This increased friction between members, and internal rivalry increased over pricing and production issues. Exceeding production quotas is an inherited problem in an oligopoly market structure model such as OPEC. Each member will try to maximize their revenue, constrained by the cost of production as well as production capacity and economic priorities. Political and social factors ultimately will affect the country's production horizon. Such differences between members in their internal political and economic problems created some internal rivalry.

The external political environment and conditions which surround member countries have also affected each OPEC member's behavior and caused rivalry among the members. However, the oil market changed after the 1967 and 1973 Arab–Israeli Wars in ways which made oil a weapon and became much more powerful. Rapid demand growth consumed US spare oil production capacity, and by 1970, net oil imports to the United States were rising rapidly.

The United States is no longer the source of spare capacity and a "security margin" in the oil market. At the same time, oil production in the Middle East also grew quickly, meeting two-thirds of global demand growth between 1960 and 1970.

Two major events in the 1970s shaped OPEC behavior as well as the political and economic future of individual members. The first was the Arab–Israeli War in October 1973, while the second event was the Iranian Revolution in 1979. These

TABLE 2.1

Selected Annual Average US Crude Oil Price (1950–2020) ($/Barrel)

Year	Nominal Price	Real Price
1950	2.77	29.70
1970	3.39	22.54
1973	4.75	27.37
1974	9.35	48.84
1980	37.42	117.30
1986	14.44	33.97
1990–2000	18.50	32.30
2005–2014	73.95	87.20
2020	50.96	50.96

Source: Illinois Oil and Gas Association

two events divided OPEC; Arab member countries in 1973 put an oil embargo on oil export to those countries who supported Israel, mainly the United States.

This caused a sharp increase in oil prices and caused a world oil crisis, especially in the United States. The real oil prices went up by about 80 percent: from $27 per barrel in 1973 to almost $49 per barrel in 1974, as presented in Table 2.1.

However, non-Arab members such as Iran and Venezuela took advantage of this embargo to raise their production in order to substitute for the supply loss, even though the embargo was wrongly understood as orchestrated by OPEC embargo and therefore blamed for it in the international media.

In 1979, social unrest in Iran sparked the Iranian Revolution, which caused oil disruption from Iran. This was the second major oil crisis of higher oil prices and a cut in oil supply.

As in the previous crisis of 1973, some countries with excess production capacities were able to substitute for the loss in oil supply from Iran. This incident aggravated the internal rivalry among OPEC members.

Each country was striving to pump more oil into the market hence taking advantage of the high oil prices, which reached more than $80 per barrel in real terms. Furthermore, the decade of the 1980s started with the Iran–Iraq War in September 1980. During the next eight years (the duration of the war), the other OPEC members – mainly Saudi Arabia, Kuwait, and the United Arab Emirates – compensated for the oil supply interruption because of the war; they were suffering from possible destruction of their oil facilities due to the war and there was a fear of closing the Strait of Hormuz.

The internal rivalry was concentrated on internal administrative matters, such as the appointment of the Secretary General and on pricing issues such as whether to have higher oil prices as Iran advocated, or lower oil prices as the Saudis wanted. This was the beginning of a long-term rivalry that would separate the Hawks from the Doves and divide OPEC for future years.

This issue, as it turned out, depended on how much oil OPEC could produce; this was in the hands of Saudi Arabia as the major oil producer or swing producer. This game of production manipulation continued until the market was saturated by an oil glut which caused oil prices to collapse and reach as low as $9 per barrel in 1986. The conflict between OPEC members over oil prices and production levels continued until the Iraqi invasion of Kuwait in August 1990. This disrupted oil production from both Iraq and Kuwait with losses of almost 5mbd.

Due to supply security and preventing oil prices from shooting up, the other members produced more oil outside the quota agreements to compensate for the oil loss. This was considered by OPEC to be an attempt to reduce market tensions beyond the usual coordination.

After the liberation of Kuwait under the "Desert Storm" operation in 1991, the country gradually returned back to a normal production rate of 1.8 mbd.

Iraq, on the other hand, was placed under sanctions which continued until the invasion of Iraq in 2003. The invasion coalition forces were led by the United States to topple Saddam Hussein. During the Gulf Wars in the 1980s and 1990s, OPEC members were competing on production quotas which were facilitated by the absence of Iraq and Iran during the Iran–Iraq War from 1980 to 1988, and the absence of Iraq and Kuwait in 1991 until Kuwait resumed its production; nevertheless, sanctions continued against Iraq during the 1990s.

Subsequently, the internal rivalry among members was usually over the production allocations and to a certain degree on oil pricing between Saudi Arabia and Iran. This rivalry between the two countries boiled down to their difference in political and social agendas and how influential each country is in affecting the politics of the Middle East in particular – and the Islamic world at large.

2.2.1 DOVES VERSUS HAWKS

While OPEC was gaining power as an oil market key player in the 1970s, the largest producers became more influential. Saudi Arabia, Iran, and Iraq are examples. Other members played an influential role, given their production capabilities. It was well constituted that all members are net oil exporters with equal votes, but how much they produce and export depends on their oil reserves, production capacities, their production costs, and political and economic priorities.

Based on these criterions, two camps emerged: the Doves and the Hawks. Of course, the Doves are those countries with high reserves, low production costs, and high production rate capacity. These countries include Saudi Arabia, Kuwait, the UAE, and to a certain degree Iraq, especially during its war with Iran.

Regardless of their political motives, the Doves advocate a low oil price and high output. This is to increase their oil revenues but not necessary maximize their profits. As far as political incentives are concerned, the Gulf countries – as well as Iraq – have the desire to put economic pressure on Iran by flooding the market and obtaining lower oil prices in order to reduce the oil revenue of Iran. This movement towards low oil prices was also encouraged by the US and Saudi Arabia to hurt the former

Soviet Union in the late 1970s and early 1980s during their war in Afghanistan. This was in order to limit the income from hard currencies.

Non-OPEC counties including the United States, United Kingdom, Norway, and Russia were also added to the problem of oil glut. On the other hand, the countries who were under the Hawks' flag were those countries with low oil reserves and a high cost of production, which implies lower oil production and high income to satisfy their present time preference.

Examples of oil price hawks from OPEC countries are: Algeria, Libya, Venezuela, and Iran. With respect to Iran and Libya, their political motives are essential factors in taking this position. In the case of Algeria, the limited oil reserve, and social and economic conditions are factors in their being an oil price hawk.

The idea of Doves versus Hawks became very clear after the first two decades of the foundation of OPEC. That was when demand and supply metrics determined oil prices, and each OPEC member had to protect its own vested interests. In the 1990s, it was noted that Saudi Arabia was leading the Doves and expecting to deal with the Hawks (mainly Iran) in discussing OPEC output and prices both behind doors and openly in press conferences.

Sometimes, Venezuela and the OPEC Secretary General played as mediators throughout negotiations between the two countries. An important case of Doves versus Hawks was in a June 2011 ministerial meeting, which took place in Vienna.

In that meeting the ministers failed to agree on a production increase, in spite of the United States and Europe's hint to level up production from 24.8 mbd to the actual level of 26.3 mbd, in a way to reduce oil prices from around $100 a barrel for Brent crude. After lengthy discussion, the outcome of that meeting was a victory for the Hawks and a loss for the Doves led by Saudi Arabia (Defterios, 2011).

In most cases, Iran as a price hawk is a hard negotiator which applies the strategy of inflating its actual reported output and pretending that it works for price reduction but follows the inherited behavior of most OPEC members in violating any assigned quota.

Most recently, as has been seen, OPEC's Hawks became Doves and vice versa – as in the case of Saudi Arabia which turned out to be a Hawk in reducing output to have higher prices. This is due to changes in geopolitics, political interests, and the newly adopted economic strategies of the member countries.

2.2.2 SAUDI ARABIA VERSUS OPEC

Saudi Arabia is one of the five founding members of OPEC. Saudi Arabia considers OPEC a shelter against the power of the major international oil companies or countries.

Between the aggressiveness of Venezuela and Iran, Saudi Arabia is a moderate in dealing with oil companies. Saudi Arabia admitted its limitations with technical and commercial aspects. Therefore, it tried to convince other members to maintain workable and reasonable relations with international oil companies. In the 1960s, OPEC – including Saudi Arabia – from time to time placed its claims for price adjustment to be considered by oil companies under the OPEC umbrella. Saudi Arabia, like the rest of the member countries, asked oil companies to sit down and renegotiate the terms and conditions of the concession agreements between them and the corresponding oil companies which were operating on their soil.

Given the current and potential massive oil reserve of Saudi Arabia, it became de facto leader within OPEC in terms of price and production decisions.

This in addition to its political power which is attributed to its Islamic role as custodian of holy mosques which are invested in by Saudi Arabia in building its image of spreading international peace and influencing oil market stability and the welfare of the world economy.

The moderate role of Saudi Arabia in OPEC gave the organization bigger room to penetrate an oil industry highly controlled by the major international oil companies. However, over the years and with an increasing number of member countries, Saudi Arabia found it more difficult to manipulate OPEC. This difficulty has become clearer with political differences and conflicts between members as in the cases between Iraq and Kuwait, Iraq and Iran, and recently between Saudi Arabia itself and Iran.

Conflict of interests became an issue within OPEC, which made Saudi Arabia shy away from taking the leading role in OPEC. Nevertheless, sometimes, the Saudis use OPEC as a vehicle to pass on their oil policies. However, other rival countries such as Iran or Libya stand against it. Now more than ever, consensus among OPEC members is not easy, nor has it ever been.

Sometimes, there comes the situation where Saudi Arabia considers leaving OPEC. This turns out to be a matter of cost/benefit analysis when seeing what could happen if Saudi Arabia walked away, or left OPEC. The implications have to be studied and analyzed from both sides, i.e. Saudi Arabia and OPEC: What might Saudi Arabia gain and what might OPEC lose? Certainly, both would lose: OPEC would lose its key leader (who would carry the flag?) and Saudi Arabia would lose the umbrella, which allows Saudi Arabia to pass on its policies and oil interests to the rest of the world.

The main question, which determines if the OPEC and Saudi Arabian relationship remains, is who leads who. The general impression is that Saudi Arabia has dominated OPEC for the last six decades. In many cases, Saudi Arabia has to play down its leading role in a way to avoid public criticism and hide behind OPEC consensus if the decision is in the interest of both.

More recently, there have been a few arguments from analysts and government officials saying that Saudi Arabia should withdraw from OPEC.

These arguments stand on the basis of the idea that OPEC is starting to lose its power to regulate the world oil market and manage the tensions between members, especially between the two largest producers, Saudi Arabia and Iran. Furthermore, even though Saudi Arabia is an oil producer, it is starting to behave like the consuming countries who see less importance in oil – at least in the medium- to long-term – and more tendency to depend on renewable energy.

Some argue that with world economic change, Saudi Arabia is now in a position to act freely with regard to its oil policy and does not need a cover to act against the weak OPEC.

This might give the impression that Saudi Arabia is thinking of leaving OPEC. In the past, a few members have withdrawn from OPEC, such as Ecuador, Indonesia, Gabon, and most recently Qatar. However, Ecuador rejoined after its suspension in 1992, while Gabon rejoined in 2016 from its leaving OPEC in 1995 while Qatar left OPEC on January 1, 2019.

When it comes to Saudi Arabia, the decision to pull out of OPEC is not easy, given the position of Saudi Arabia as a founding member and a major OPEC producer.

Saudi Arabia finds staying in OPEC to be an opportunity to control the organization and pass on its oil polices to most of the members. This would be leverage for Saudi Arabia in the world oil market as well as political bargaining power inside and outside OPEC. But life is not always perfect – there are some OPEC members who are influential – Iran and Venezuela are examples who, in most cases, stand against Saudi Arabia to advocate their own interests. This saying could be applied to other members but with less degree of effect and influence.

Saudi Arabia apparently has some obligations toward holding OPEC together and keeping the world oil market sustainable and stable in spite of the declining importance of OPEC and oil in the medium- to long-term. This implies that Saudi Arabia would think twice before deciding to leave OPEC.

2.3 EXTERNAL CONFLICTS

2.3.1 OPEC VERSUS THE WORLD

The world sees OPEC as an economic international organization that deals with oil and controls the world oil market with two-thirds of the world oil reserve. In the early days of the foundation of OPEC, the world did not pay much attention. This is because oil as a commodity was not widely used all over the world like now, and the strategic feature of oil was still not recognized.

The establishment of OPEC was just a reaction from five less developed countries that happened to be major oil producers in the face of greediness of the major international oil companies. The voice of OPEC to have a better relationship with oil companies and more fair oil prices was not listened to.

With the oil crisis of 1973, the world began to recognize the presence of OPEC and its importance to manage the world oil market. As a matter of fact, the world put the blame on OPEC for the increase in oil prices. As a result, in 1974 the industrialized countries established IEA as an oil-consuming country bloc against OPEC. Its announced objective is to monitor oil stocks in its member states to the level of at least 90 days of oil net imports. In order to balance oil demand and supply, IEA has to either build up or withdraw the oil stocks as a complementary effort to OPEC's objectives of reaching a stable oil price. A by-product of such a strategic mission, IEA works as a source and base of data and information on the oil market and the economic conditions of its members.

The western media bias against OPEC has created stereotypical images of Arabs and OPEC as a threat to the world economy.

The 1970s can be considered as the golden age of OPEC for its achievement in setting oil prices and from a press perspective, OPEC received wide coverage by the

FIGURE 2.1 Saudi oil minister Mr. Ahmed Zaki Yamani. Source: OPEC PICL-1973 Portraiture(25)JPG, www.opec.org

press and other modes of media all over the world. It became known even to the layman; oil ministers of producing countries, on the other hand, became celebrities like movie stars or famous football players. One clear example was the Saudi oil minister Ahmed Zaki Yamani, Figure 2.1.

He served as the Saudi oil minister from 1962 to 1986, and played a key role during the Arab oil embargo of 1973. He received an international reputation and wide media coverage on December 21, 1975 when he and other OPEC oil ministers were taken hostage – in Vienna during the ministerial meeting at the OPEC headquarters – by six persons led by the Venezuelan terrorist Carlos "the Jackal" under what they called the "Arm of the Arab Revolution" group, with the claim of working for the Liberation of Palestine.

The terrorists were determined to execute Yamani and the Iranian oil minister Amuzegar. It was stated by an OPEC official that Carlos entered the conference room after killing two Austrian police officers and an Iraqi security guard, in addition to a Libyan economist.

A terrorist started shooting the ceiling while the ministers and other delegates hid under the tables. After identifying Yamani, the hostages were sorted into groups.

After negations with the Austrian authorities the terrorists decided to fly to Algeria where the story ended and oil ministers released (Weiss, 2004). There are conflicting stories as to who was behind this operation and who funded it and who paid the ransoms. Some stories said Palestine commanded it, others mentioned Libya or Iraq. Yamani was dismissed by King Fahad of Saudi Arabia in 1986 when the oil price collapsed as a result of adopting the role of "swing producer" by Saudi Arabia where Yamani was advocating that policy.

Yamani was known for his charismatic personality and he was an outspoken man with shrewd political skills. This is due partly to his cultural background as being born in the Mecca region and his father and grandfather were scholars of Islamic law in addition to his educational background as a lawyer.

Throughout the 1970s OPEC continued in the glare of the media – and for the following decades – but not with the same momentum or intensity. The world started to understand the real role of OPEC, i.e. that it is not a weapon in the hands of its members to be pointed at the world when they want, but rather OPEC is an oil market regulator seeking market stability and fair oil prices.

This appeared clearly during the oil price collapse in the mid-1980s as a result of oil oversupply; during the 1990s because of the economic crisis of the Asian Tigers; and more recently during the world financial crisis of 2008.

2.3.2 OPEC versus Non-OPEC

The notion "non-OPEC" was not invented when OPEC was first founded. At that time, there were major oil producers, mainly the United States and the former Soviet Union, and a few small producers scattered all over the world. Apart from that, there were the Middle Eastern producers with a high reserve and production potential which formed OPEC. However, the issue of non-OPEC oil production emerged in the 1980s after exploring and developing new areas as a result of the 1973 and 1979 price shocks. High oil prices encouraged exploration and development in high-cost areas such as the North Sea and Alaska North slope. New oil producers evolved, such as Norway, UK, Mexico, Canada, and Oman.

Those countries among others constituted what is referred to as "non-OPEC" oil producers, which became competitors to OPEC and stated they wanted to be a part of the market share.

Non-OPEC producers collectively or individually became effective players in balancing the world oil market. The non-OPEC role became so clear during the oil glut of the 1980s when supply exceeded demand.

Since then, it is recognized that managing the oil market is no longer the sole responsibility of OPEC.

It is most essential to have a close and effective coordination between OPEC and non-OPEC countries in balancing supply with demand. Over the years, OPEC has extended its efforts to bring non-OPEC into its camp, in order to unify their production decisions or to maintain non-OPEC support for OPEC output. OPEC invited key non-OPEC producers to its conference meetings as observers, and occasionally both groups organized official common meetings, which were held in non-OPEC countries. One of the important meetings was held in Oman in 1993 where 14 non-OPEC oil-producing countries were invited to put the issue of coordination into effect.

In any case, the outcomes of these meetings and the agreements which were reached between the two parties remain but are not obligatory. In any case, the commitment of non-OPEC countries to control their output levels in order to support OPEC production decisions do not go in most cases beyond lip service.

Apparently, the joint agreements between OPEC and non-OPEC failed and did not materialize because non-OPEC naturally consider themselves free riders.

However, bilateral understanding or gentlemen's agreements take place from time to time between one member from OPEC and a counterpart from non-OPEC; for example, between Saudi Arabia and Russia.

As a major non-OPEC producer, Russian relations with OPEC date back to the early days of OPEC before the collapse of the Soviet Union in 1991 and continue to this day. Russia aims to have high oil prices and capitalize on its previous political relations with the OPEC price hawks Iraq and Algeria, in spite of its reluctance to reduce their oil exports.

Recently, the political influence of Russia has been increasing in the Middle East and its political relationship with the major oil exporter Saudi Arabia has improved. Therefore, it is to be expected that there is more coordination in oil production policy between the two countries in particular – and between Russia and OPEC in general. There is a question as to why Russia is not a member of OPEC, even though it is an oil exporter.

However, there were some reports indicating that OPEC was approached by Russia to join, but because of the nature of the Russian oil industry – it is privatized and strategic to the Russian government and contributes more than 50 percent to its budget as taxes – Russia would rather be independent in its oil production decision from being under OPEC control (Smith, 2017).

Regarding non-OPEC producers, they lack a formal collective commitment toward controlling their oil supply, which comes mainly from the United States, Russia, Norway, Mexico, Brazil, and Malaysia. This heterogeneous group makes it more difficult to coordinate oil production among themselves – rather than between them and OPEC.

However, both OPEC and non-OPEC producers recognize the importance of constructive coordination and dialogue in order to maximize their common benefits, particularly during critical situations of market instability. In the 1990s, the coordination among different beneficiary parties – including the consumers – was extended in a formal producer–consumer dialogue.

Over the last three decades, non-OPEC oil supply has expanded and became a major factor in the world's supply/demand balance. This can be seen by the decline in OPEC's market share in world oil supply and demand.

Table 2.2 shows the share of non-OPEC production in total world oil production from 1960–2018, as compared to OPEC's production. The increase in non-OPEC production has been caused by the increase in world demand as well as due to better terms of production deals between international oil companies and oil-producing countries outside OPEC, in addition to the latest developments in technology.

From an economic point of view, it was clear that the prevailing oil prices at $20 barrel in the 1980s and 1990s were acceptable to allow for more oil production, given the exploration and development costs of around $7 per barrel (Alsahlawi, 1993).

TABLE 2.2

Non-OPEC Production Share in World Oil Supply (1960–2018)

	Non-OPEC production (mbd)	Non-OPEC % of world supply
1960	13.4	60.3
1990	35.3	62.0
2018	47.5	57.6

Source: OPEC statistics

2.3.3 OPEC VERSUS THE US

Before OPEC, the US – as the world's largest oil producer – was controlling oil prices through its major oil companies and oil trading system. Eventually the oil market started to be controlled by OPEC after its establishment while the role of the US diminished as a key leading oil producer and a direct influence on the oil market.

However, the United States remains a major oil producer and the largest oil importer where most of its oil imports come from OPEC.

Due to the oil crisis of 1973, when Arab oil producers banned exporting oil to the United States because of its support of Israel during the Arab–Israeli War, oil prices sharply increased. The nature of the OPEC–US relationship boils down to the US relationship with each member county and US interests in different regions.

In terms of oil business, US relations are stronger with Saudi Arabia and Venezuela where US oil imports from these two countries constitute more than 20 percent of its total oil consumption. How far OPEC might affect US energy and oil policy depends on the international economic and political events and circumstances. The US's different administrations have a certain impact on US energy policy and on its relationship with OPEC. This appears clearly, for example, during the Nixon, Reagan, and Bush Senior administrations.

US wars and peace have certain effects on its demand for oil and how much it imports from OPEC countries and how much the US might manipulate the global oil market or OPEC production decisions.

This appeared clearly in the Vietnam War and the Middle East Wars during the liberation of Kuwait and invasion of Iraq. More recently, US–Iran tensions over the Iran nuclear program, when the US put sanctions on Iran imports and exports of oil to prevent uranium enrichment.

Following 2008's high oil prices, shale oil development and production in the US and other region started and gained a high momentum. The US has the largest reserve and production, where its production from shale oil reached 8 mbd in 2019 according to the US energy information department. This put the US on top of the list of oil producers, with about 12 mbd, which exceeds the productions of Saudi Arabia and Russia.

The technology of hydraulic fracturing is currently used in producing shale oil and gas has been developed greatly in the last ten years. Over the last few years, the cost of shale petroleum production has decreased substantially. It became economically viable to produce at an oil price of less than $40. If this trend continues, the US will become a net oil exporter in the next decade.

Such developments will put the US in competition with OPEC over the market share in world oil market.

Because of the increase of US shale oil production and its impact on its total production, OPEC market share has shrunk to almost 30 percent. There are more talks between OPEC and Russia and other non-OPEC countries to manage their production in a way to sustain oil prices but this depends on how far each country sticks to its word.

An important dimension of the relationship between OPEC and the United States is the link between the US dollar and the price of oil, where oil trade is denominated in the US dollar. One implication of such a link is the effect of the change in dollar value on the price of oil and ultimately on oil revenues of OPEC countries.

As a combined effect of an increase in oil demand and consequently in oil price, this will lead to increase in the demand for the US dollar, which results in a higher dollar price compared to other currencies.

During the first decade after the establishment of OPEC, oil prices were at low levels and the exchange rate of the dollar was fixed based on the Bretton Woods system. However, with the fall of the dollar value compared to the Special Drawing Rights (SDRs) of the International Monetary Fund (IMF), the Bretton Woods system was modified to a flexible exchange rate system which was adopted by the United States (Alsahlawi, 1994–1995).

From 1970 to 1979, the decline in the dollar's value continued, while the oil price increased over the period 1970–1981 including the sharp price increase of the supply shocks of 1973–1974 and 1979–1981. The increase in demand strengthened, the increase in oil price. This was reflected in the world oil export earnings, which were $250 billion in 1981 and transferred from oil-consuming countries to oil-producing countries. The majority of dollar earnings of producing countries are spent on imports of goods and services, and the remaining invested in US financial markets. Such capital movements are known as movements of "petrodollars". The whole cycle of the dollar is benefiting the US economy and its political power, which means a lot to the United States.

Since 1980s, the oil market has been characterized by fluctuations in oil prices and at the same time variation in the value of the US dollar.

Such volatility in oil and financial markets raised the question whether to continue pricing oil by the dollar or switch to another currency.

This of course has implications for the US political and economic power, not only within OPEC or its members but for the world at large. This became a concern for OPEC member countries, because a depreciation of the dollar would reduce the purchasing power of a barrel of oil.

The issue of an alternative oil-pricing currency became a subject of debate in OPEC and outside OPEC. Each interest group looks at the matter seeking its own benefits.

The United States for example, pushes to keep the status quo, while each OPEC member assesses the impacts on its trade patterns, oil revenues, and bilateral relations with the United States or with other countries. The European Union were advocating using the Euro as the oil currency. Yet any alternative currency is open to political and economic criticism, such as the realities of trade relationships and possible trade alliances with China and India (Alsahlawi, 1997).

Meanwhile Iran, for political reasons, encourages moving away from the dollar to retaliate against the United States, yet any pricing scheme for oil using the US dollar or other currency or a basket of currencies will have certain strategic and trade implications – especially when it comes to the United States, which has important allies in the form of OPEC members such as Saudi Arabia, Kuwait, and the UAE.

The general impact is not limited to strategic relations with the United States. It extends to the foreign reserves of OPEC countries as well as international trade balances (Alsahlawi, 2009). The economic growth and inflation are not immune from the variation of the US dollar and/or from oil price fluctuations because of changing the currency of oil pricing.

2.4   SAUDI ARABIA VERSUS THE US

Saudi Arabia–United States relations in the oil business started in 1933 when a sixty-year concession agreement was signed between the Saudi government and Standard Oil of California (SOCAL), an American oil company (now Chevron), to explore for oil in the eastern area of Saudi Arabia. Figure 2.2 shows the first Saudi Finance Minister, Abdullah Al-Sulaiman signing the concession agreement from the Saudi side and Mr. Lloyd Hamilton from the American oil company SOCAL.

The first commercial discovery was in 1938, with high potential of oil production until it reached a high peak of 12 mbd in March 2015. That figure was based on a statement made by Saudi oil minister Ali Al-Naimi. In 1936, SOCAL and the Texas Oil Company (Texaco) formed a partnership in Saudi Arabia and founded the

FIGURE 2.2   Signing the concession agreement between Saudi Arabia and Standard Oil of California (SOCAL). Source: https://archieve.aramcoworld.com

Arabian-American Oil Company (Aramco) in 1944. This consortium was expanded to include what would later be known as Exxon and Mobil; the Saudi government gradually bought out the foreign shareholders beginning in 1980.

When Saudi ownership reached 100 percent, Saudi Aramco was established in 1988. Saudi–US oil relations continue, where many technical operations are still handled by American companies. For example, Chevron, Dow Chemical, and Exxon Mobil have joint ventures in Saudi Arabia in the refining and petrochemical sectors.

On the trading and marketing side, Saudi Arabia exports around 1.2 mbd to the United States, which ranks Saudi Arabia as one of the largest oil exporters to the US. However, US oil production lately surpassed Saudi oil production, thanks to rapid US development in shale oil production which made the US an oil exporter and a potential competitor to Saudi Arabia. Going back to the early concession agreement which set the path of the oil relations between Saudi Arabia and the United States, article 1 of the agreement provided that

> The Government hereby grants to the Company on the terms and conditions herein-after mentioned, and with respect to the area defined, the exclusive right, for a period of 60 years from the effective date hereof to explore, prospect, drill for, extract, treat, manufacture, transport, deal and export. This right does not include the exclusive right to sell crude or refined products within Saudi Arabia.

It worth mentioning that over the concession years, there were several disputes between the government and the partners on exporting, selling, and transporting oil. The disputes are always settled on the basis of international law, and state sovereignty interpretation.

After several amendments of the concession agreement, the most recent amendment – which is effective as of January 1, 2020 – gave Saudi Aramco the exclusive right for oil and gas productions. The royalties from crude oil and condensate output will be reduced to 15 percent from 20 percent. The amendments include an increase of marginal royalty rates to 45 percent from 40 percent on Brent prices above $70/b up to $100/b and an increase in the marginal royalty rate to 80 percent from 50 percent on Brent prices above $100/b. The government of Saudi Arabia will compensate Saudi Aramco for extra carrying costs for the government, setting prices on oil products and gas. Furthermore, the Saudi government will reduce the tax rate on downstream business. These changes on the concession agreement aimed to put Saudi Aramco up for an initial public offering (IPO).

Recognizing the strategic and massive Saudi oil reserves, and based on the experience of American oil companies in dealing with the Saudi government, the United States decided to further formalize its political and trade relations with Saudi Arabia in a way to enhance both countries' interests in oil supply security and oil price stability. As a result, US President Franklin D. Roosevelt (FDR) decided to meet officially with King Abdulaziz, the founder of the kingdom of Saudi Arabia, onboard the USS Quincy in the Suez Canal in Great Bitter Lake, Egypt on February 14, 1945. This was a historical meeting where the first time a US President met a Saudi King and set the reference line for the Saudi–US relations which is referred to as the

FIGURE 2.3   Quincy Pact. Sources: https://www.history.com/news/fdr-saudi-arabia-king-oil

"Quincy Pact", Figure 2.3. For the United States, the supply of Saudi oil is guaranteed and in return the US committed to protect Saudi Arabia and consider it as an important ally in the region.

Building on the "Quincy Pact", many subsequent agreements on different political, trade, finance, and military issues have been signed between Saudi Arabia and the United States.

Based on the special relations between both countries, Saudi Arabia became a very important ally for the US in the Middle East. On the oil front, no one would expect significant differences between Saudi Arabia and the US especially when it came to oil prices or Saudi Arabia's leading role in OPEC with regard to oil production. Saudi oil policy is starting to take an independent position, when it comes to exporting its oil to emerging economies such as China and India.

Saudi went even further by investing in the downstream of oil industry in those countries. To diversify its oil interests, with OPEC losing its power, it is expected that Saudi Arabia will face its differences with the US openly without working under OPEC coverage as before.

Apparently, there is a close coordination between Saudi Arabia and the United States regarding what the range of oil prices should be. However, the US cannot dictate an oil price, but certainly, the Saudi Arabia uses the US-preferred oil prices as a hint that may be mentioned during the negotiations among OPEC members.

In the past, there was conflict between the two countries over the Arab oil embargo which was imposed by Saudi Arabia and other Arab oil-exporting countries on the US and its allies for supporting Israel during the 1973 war between Arab states and Israel. After that the relationship between the United States and Saudi Arabia was in harmony, especially during the Soviet War in Afghanistan in the 1980s, when Saudi Arabia orchestrated low oil prices. This was because of the price war between Saudi Arabia – representing OPEC (except Iran) – and non-OPEC countries. There was a slight conflict between Saudi Arabia and the US when the Saudis were over-producing in 2014 to hurt the high-cost US shale oil producers by lowering the oil price. This also has been repeated recently due to low demand because of the coronavirus epidemic. Saudi Arabia went alone into a price war with Russia, which caused low oil prices that in turn affected US oil producers and exporters. However, with more coordination between Saudi Arabia and the United States, the market responded positively as Saudi Arabia and OPEC+ reduced their output.

However, Saudi Arabia generally maintains an adequate oil production to fulfil the world's needs from oil and to meet US oil demands. While the United States is guaranteed to be supplied by Saudi oil, the US kept its obligations to protect the route of oil shipments from the Gulf to the markets.

PART III
Energy Technology and OPEC

The capital investment which is needed in developing technologies and spent on renewable energy vary among different types of energy, either solar, wind, or hydropower. Investment in technology is also needed for developing and producing unconventional oil and gas, especially when oil prices are high, in order to cover the high production costs of shale oil and gas.

3.1 ENERGY AND TECHNOLOGY

Technology, now more than ever, is expected to shape the future of OPEC in terms of energy supply and demand. Nevertheless, technology will continue to magnify and diversify energy supply in order to meet the ever-increasing demand for energy. It is important to recognize the fact that because of technology, the cost of supplying renewable energy has fallen considerably.

In the medium term, the cost of supplying alternative energy is estimated to be 30 percent less than current levels. The extraction of oil and gas has also benefited from the cost-saving resulting from technological advances.

This can be seen in the surge of tight oil and shale gas production in the United States. Table 3.1 presents the share of different fuels in the world's energy supply, with the projection of share until the year 2030 (Alsahlawi, 1994).

It is noted that the supply of renewable energy has been increasing over the years. This is due to the change in energy demand structure and technological advances. As a secondary energy supply, nuclear and renewable energy are used in power generation. Clean energy constitutes more than 20 percent of the total energy supply and will produce more than 40 percent of electricity by the year 2030. Advanced coal and gas technologies, such as coal-fired generating plants will be used to generate another 50 percent of power in 2030, with improving energy efficiency and less emissions.

From an energy demand perspective, energy consumption is determined mainly by the price of energy and economic growth among other factors, as indicated in Figure 3.1.

The price decrease increases the demand: this is basic economics. Since the price is negatively related to the quantity demanded while income increases, as explained by economic growth and population growth, this will increase the demand.

There are other factors, such as fiscal and monetary policies, in addition to policies regarding the preservation of the environment and managing the demand for energy. These have mixed effects on energy consumption.

However, all factors that determine energy consumption are closely interrelated and affect each other. For example, technological developments affect energy demand and supply in many directions.

TABLE 3.1

World Energy Supply Mix (Fuel Shares), 1970–2030

Energy Types	1970	2010	2030
Oil	45.5	34.0	39.0
Gas	18.0	21.0	20.5
Coal	25.0	26.5	20.5
Nuclear	0.7	5.0	5.5
Renewable Energy	11.3	13.5	15.5

Source: Author's estimates

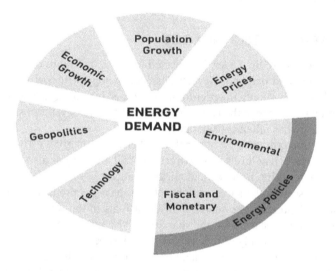

FIGURE 3.1 Energy demand factors.

First, this is through its direct effect on demand by reducing energy use and improving energy efficiency.

Second, technology affects energy security by expanding the supply reliability and developing alternative energies. Third, there is an indirect effect through energy prices, which are expected to fall with more supply and better energy-saving technologies. This might have the reverse effect by increasing energy consumption as prices decrease. However, with imposing environmental measures and enforcing energy conservation, demand will be controlled. Furthermore, technology plays a crucial role in reducing pollution and controlling carbon emissions.

As a matter of fact, technology plays an essential role in the energy supply as well. Several studies have attempted to analyze the impact of technology on the supply and demand for energy in different sectors, such as the industrial sector (Popp, 2001).

It has been found that innovation and technology change the demand and supply picture for energy.

From the supply side, technological developments over the last four decades have sparked the evolution of a wide range of alternative renewable energies, as well as conventional and non-conventional energy types.

Nevertheless, technology has improved oil and gas exploration and production methods and also reduced their cost.

As far as electricity is concerned, new technologies and advanced systems such as co-generation and hybrid power systems have been used with different forms of renewable energy to generate electricity.

Generally speaking, since the energy crisis of 1973, technology has become a crucial factor in achieving energy security and managing the demand for energy. On the demand side, technology has helped to reduce energy use by means of several conservation policies.

However, conservation will not be met unless energy efficiency requirements and standards are applied. It has become clear that energy efficiency will lead to general economic efficiency and better use of energy at lower prices. One common and important measure of energy efficiency is energy intensity, which is the energy use per dollar value of output (Alsahlawi, 2013).

3.2 EFFECTS OF TECHNOLOGY

Energy and technology are two sides of the same coin. Both interact with each other and form what might be considered two-sided markets or networks. The evolution and integration of energy with technology at a wide level began in the middle of the last century.

Such integration reshaped the world's economic systems into a new order. Needless to say, energy is a key factor in the technological developments of the past and future.

The interdependence between energy and technology appears more clearly when it comes to the role of technology in energy development. Since World War II, energy production and consumption have increased rapidly at different rates. Figure 3.2

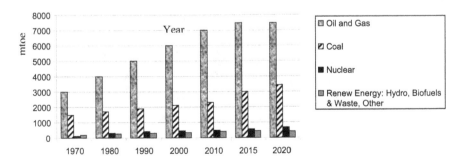

FIGURE 3.2 World energy supply by fuel, 1970–2020 (Mtoe). Source: IEA World Energy Statistics, and author's estimates.

shows that the total energy supply increased by more than 3 percent a year from 1970–2020.

The effects of technology on energy supply were captured after the energy crisis of 1973. It was noted that more nuclear and renewable energy types were developed. Furthermore, new oil exploration and production activities from high-cost areas such as the North Sea and Alaska increased as a result of the high oil prices.

As far as energy consumption is concerned, there was a rise in energy consumption following World War II.

This was caused by the need to rebuild the destroyed infrastructure in different parts of the world after the war.

Moreover, the ensuing relative political stability and economic prosperity in the United States and Europe boosted energy consumption.

The increase in energy consumption became a trend which continued until the first oil price shock of 1973, and the second oil price shock because of the Iranian Revolution in 1979.

As shown in Figure 3.3, world energy consumption rose moderately by 2.0 percent annually over the period from 1970 to 2020. The modest growth rate in energy consumption compared to the rate of energy production was due to the implementation of energy conservation measures with the utilization of technology.

Development of technology has facilitated the application of energy conservation policies and allowed substitution between different energy types. It is noted that, because of technology, energy consumption patterns have changed worldwide and even within different economic sectors.

Energy and technology are systematically linked with large impacts on environment, economic performance, and energy prices. It is an ongoing debate as to whether negative environmental impacts are more associated with energy production or consumption.

As energy use increases as a result of the growth in world population and expansion of economic activities, the environment is expected to be affected negatively.

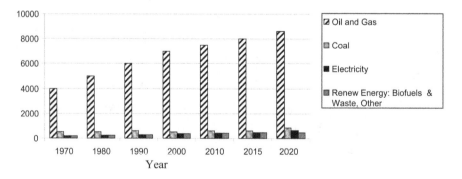

FIGURE 3.3 World energy consumption by fuel, 1970–2020 (Mtoe). Source: IEA World Energy Statistics, and author's estimates.

Climate change effects due to global warming, radiation emissions, and greenhouse gas emissions (mainly CO_2) are examples of this.

In spite of international climate change negotiations and subsequent agreements, such as the United Nations Framework on Climate Change (UNFCCC) in 1992 and the UN Climate Convention and other climate negotiations from the European Commission (Soytas and Sari, 2009), it was found that CO_2 emissions are rising at almost the same rate as world economic growth.

There is an interrelationship between energy mix and the environment: this determines the amount of CO_2 emissions. It was found that 45 percent of the total CO_2 emissions are caused by coal consumption and 50 percent from using oil and gas (Granger and Keith, 2008).

It is therefore clear that there will be major negative consequences on OPEC oil demand and its members' revenue, by applying different environmental protection policies such as energy/carbon tax for the mitigation of CO_2 emissions.

3.3 THE ECONOMICS OF ENERGY TECHNOLOGY

Energy is the power that is needed to perform an activity. Historically, human muscles were utilized as the original source for energy, and over time new natural resources were discovered and utilized – oil was one of these primary sources.

With technological developments, other forms of energy have been discovered and utilized, culminating in electricity. Technological development with respect to energy is both an endogenous and exogenous process.

There is a general state of technology, which affects the quality of life and then there are technological developments pertaining to each form of energy in terms of its production and consumption or utilization.

However, moving from one technology to another or from one source of energy to another involves economic costs as well as structural modifications, which affect both lifestyle and economic and social development.

As a matter of fact, technology is another factor which played an essential role in determining energy production and consumption in the past – and will continue to do so in the future.

Patent data and R&D expenditure are used as measurements for technology and intellectual property rights (WIPO, 2013). It was found that innovation and technology changed energy consumption and caused a major increase in the energy supply.

From the supply side, the technological developments over the last four decades sparked the evolution of a wide range of alternative renewable energies as well as conventional and non-conventional energy types in the case of oil and gas.

However, the economic evaluation of these technologies needs to be explored and implemented.

Technological improvements are obvious with respect to the supply of fossil fuels.

Recently, unconventional oil and gas exploration and development technologies have increased the supply of tight oil and shale gas, especially in the United States. This has changed the world oil market structure and reduced oil and gas prices.

From a non-renewable energy supply perspective, technologies have contributed in developing sustainable energy alternatives and new energy and power generation systems. On the other hand, technology has developed remarkably with regard to increasing energy efficiency and providing renewable energy in order to sustain energy consumption efficiently and effectively.

3.4   ENERGY TECHNOLOGY MIX AND FUTURE ENERGY

It has become clear that energy technologies affect both the supply and demand for energy. The impact of technology has resulted in lowering energy production costs and ultimately increased energy supplies, which reduce energy prices and improve energy use efficiency.

Nevertheless, energy technology has contributed towards handling environmental impacts by cutting the negative spillover effects of energy production and consumption.

These negative effects include greenhouse gas emissions and different kinds of pollution. The mix of energy technologies is presented in Table 3.2. It shows a variety of technologies that are involved in the production and consumption (supply and demand side) of different forms of energy.

TABLE 3.2

Energy Technologies Mix and the Future of Energy

Energy Sectors	Energy types	Technologies
Supply Side	Oil and gas exploration and production	• Seismic surveys and interpretation tools • Horizontal drilling multistage hydraulic fracturing • Formation of evaluation technologies • Smart completions, acid stimulation, LNG technologies
	Coal	• Gasification
	Renewal energy	PV, CPV, and CSP systems
	Solar	• Converging technologies
	Biomass	• Acid pre-treatment technologies
	Biofuels	• Electricity geothermal production
	Geothermal	• Carbon neutral and electrolysis
	Wind	• Hydrogen storage
	Hydro	• Helium 3
	Hydrogen	• Carbon plus
	Nuclear Energy	• Reactor technologies
	Power Generation	• Fuel cells, Micro power systems and smart grid • Advanced gas turbine
Demand side	Energy efficiency	• Compact electric motors • Fuel Cells • ICT, Smart Building • CO2 capture and storage

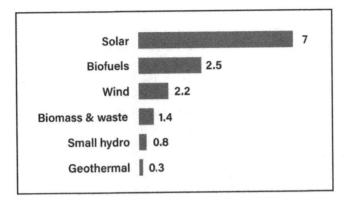

FIGURE 3.4 Global investment in renewable energy R&D by type, 2020 $BN. Source: Bloomberg, Bloomberg New Energy Finance, IEA and author's estimates.

It is important to emphasize the fact that technological development is a dynamic process. It depends on the investment cost at the R&D stage and the rate of return. The payback period is another factor that should be taken into consideration to determine the return on investment and economic viability of technology penetration within the existing systems and the cost of replacing the old devices and technologies. Figure 3.4 shows the world capital investment in R&D for each type of renewable energy.

3.5 THE ECONOMICS OF RENEWABLE ENERGY

From an energy supply perspective, technology is more associated with renewable energies rather than with non-renewables. Solar energy, wind energy, and hydropower are the most common forms of renewables.

In the case of solar energy, radiation can be collected and converted into useful energy by using the technologies of photovoltaic cells or solar thermal collectors.

For wind energy and hydro, the advanced mechanical and systems engineering are used as basic technologies to convert wind and hydro into energy. The investment cost and technological specifications are different from one type of renewable energy to another.

The technological configurations usually determine the end use of energy, whether it be heat or electricity.

For example, coal gasification technologies allow an effective production of electricity using combined gas turbines. In spite of the economic cost involved, it is noted that renewable energy technologies are developing faster than fossil fuel technologies. This is due to the practicality of utilizing available developed technologies in sectors such as aviation and energy storage.

This results in a higher output of electricity from renewable energy than electricity produced from conventional energy resources. It is estimated that renewables will produce more than 20 percent of the world's needs from electricity by the year 2030.

The increase in power generation from renewable energy is not always without a major capital cost. In general, the cost of capital investment per unit of electricity produced from renewables is higher than the cost of producing electricity from fossil fuels.

However, such costs are well justified by the attained technology advances and available utilization of these technologies, in addition to the positive environmental impacts of reducing carbon emissions. Simple cost-benefit analysis indicates that oil price is the breakeven factor between the costs and expected returns of investing in renewables. Figure 3.5 presents the world investment in renewables.

Since the recent jump of oil prices between 2007 and 2008, investment in renewable energy has increased, while the drop of oil prices in 2014 has contributed to a decline in renewables' investment. Figure 3.6 draws oil price movements presented by Brent as a benchmark in relation to world oil supply over the period 2000–2020.

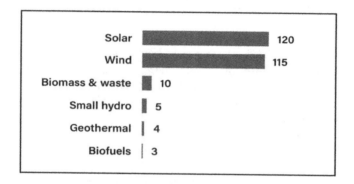

FIGURE 3.5   Global new investment in renewable energy by type, 2020 $BN. Source: UN Environment, Bloomberg New Energy Finance, and author's estimates.

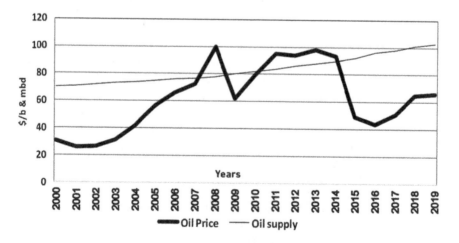

FIGURE 3.6   Oil price (Brent) versus world oil supply (mbd), 2000–2020. Source: author's estimates.

The high oil price encouraged investment in high-cost oil-producing areas and in unconventional oil and gas resources, with more investment in renewable energies. It is also noted that since 2014, the oil price share dropped as compared to oil supply, which explains the persistent surplus in the oil market in recent years. This phenomenon has magnified with the coronavirus pandemic.

As far as wind energy is concerned, it has become an important source of electricity generation. The cost of producing energy from the wind is almost 80 percent fixed cost for the turbines, foundations, road constructions, and grid connections.

The major cost of wind power project is fixed cost, while the variable costs include the operation and maintenance, land rental, and taxes. The technology of wind energy deals with improving the wind turbine technical specifications and increasing the capacity factor.

The cost of wind power per KwH produced depends on the capacity installed and the wind availability. The average capital costs for newly established wind power projects in Europe is around €1200 per KW.

In the long run – and because of technological development – the capital cost of wind energy is likely to fall to about €1000/KW as an average between onshore and offshore. From a technical point of view, wind and solar energies are variable, but are cyclical and they can be predicted depending on energy prices and the state of technology.

3.6 ECONOMICS OF UNCONVENTIONAL OIL AND GAS TECHNOLOGIES

The impacts of technology on fossil fuels' production as a result of increasing oil prices are very clear in the evolution of tight oil and shale gas production in North America. This has immense ramifications on the oil market in particular and energy markets in general, which affect the global economy and more specifically oil production.

This new oil and gas "revolution" or as often called, "shale boom", is due to technological breakthroughs in oil and gas exploration, development and production activities, such as hydraulic fracturing or, as commonly known, fracking. Technological innovations in this area have been induced by high oil prices which have overcome the high production costs of tight oil and shale gas.

This new production technique has increased oil and gas supplies and reduced oil and gas prices and ultimately discouraged further investments in developing alternative forms of energy. The prevailing excess supply of oil and gas at the current low prices has caused less consumption from other energy types, especially renewables and coal, which is the most polluting fuel.

The advances in technology have direct effects on produced quantities of shale oil and gas.

Technology is involved in almost all steps of the oil exploration and production processes. Efficient reservoir characterization and management help in extending the life of reservoirs, which leads to more production periods from these reservoirs.

Exploration technologies are used to find the reservoirs and define their boundaries, determine the target zones and expected flow rates, decide if it is feasible to produce from each zone, and evaluate the reserve quantities and the recovered ratio.

Some exploration technologies are seismic studies, advanced logging tools (imaging, fluid sampling, magnetic resonance tools), drill stem tests, and logging while drilling (LWD) techniques.

The ability to monitor the well completion's and components' integrity also contributes to extending the well's life, production, and helps avoid disaster.

Production technologies are mainly used to monitor the performance of the well (production rates), determine the oil/water or gas/oil contacts, identify the contributing zones, and inject water or chemicals into the formation to improve productivity.

Some examples of such tools are production and reservoir monitoring logging tools. There are smart completions with moving slides and packers, permanent and retrievable plugs. Wells are the most expensive and valuable asset for oil producers. The health and integrity of well components can be monitored and mitigated using advanced technologies.

The main uses of well integrity technologies are corrosion detection, leak detection, and completion damage detection.

Examples of such technologies are electromagnetic corrosion logging tools, multi-finger caliper logging tools, spectral noise leak detection tools, and down-hole video cameras. With the economic viability of unconventional oil and gas technologies, some negative environmental effects appear on the surface to be a serious matter.

First of all, there is a high possibility of contamination of the underground water by the chemicals that are used in the fracking process, in addition to the accumulation of the solid waste and destruction of the land used.

Secondly, there is the air pollution and climate change because of CO_2 and other greenhouse gas emissions as a result of using more oil and gas and less use of clean renewable energy with less efficient use of energy.

It is worth mentioning that energy efficiency technologies are costly from an energy consumption perspective and it is expected there will be less investment in such technologies given the low price of fossil fuels. It is noted that low oil and gas prices have relaxed to a certain degree the energy conservation policies and regulations.

The high cost of producing unconventional oil and gas is attributed to the expenses involved in horizontal drilling and hydraulic fracturing operations.

Records show that the horizontal well drilling cost is four times the cost of drilling a well vertically, which is estimated to be around four million dollars.

The large amount of water needed in the hydraulic fracturing and the treatment of wastewater is another cost. In the United States, the leading country in tight oil and shale gas production, the tight oil production cost ranges between $30 and $70 per barrel, while shale gas costs around $5 per MMBTU.

According to EIA, unconventional oil and gas production is expected to grow by four percent annually. Production comes mostly from North America with some production from China, Australia, and likely in the future from Saudi Arabia.

3.7   ECONOMICS OF ENERGY DEMAND TECHNOLOGIES

Several economic sectors are consuming different types of energy, either fossil fuels or renewables. Two important sectors are the main consumers for energy, electricity's generation, and transportation. However, other sectors such as industrial – other than power generation and commercial – are still important. The economics of technology from an energy consumption perspective, means basically what kind of energy technology is to be adapted in order to achieve an efficient and economical use of energy.

Yet, investing in energy efficiency depends on the state of technology and how much funds are ready to be allocated for research and development.

As a matter of fact, energy prices, particularly oil prices, play an important role in determining the level of energy consumption; meanwhile, the rate of return on capital of investment in energy efficiency projects determines the technology choice and type of energy.

Nevertheless, the effect of technology on energy consumption to improve efficiency will reduce energy usage and ultimately reduce energy prices.

It is important to point out that any efficiency gain will be also reflected on preserving the environment and meeting the objectives of climate change. Nevertheless, it is most essential from an economic point of view to have cost-effective technologies.

The real technological challenge in energy consumption is to develop low carbon technologies and sustainable energy systems in all consuming energy sectors. Another challenge is to save energy through meeting energy efficiency and conservation targets.

In the transportation sector, technology has contributed to the improvement of the efficiency of vehicles to meet fuel economy standards.

The automobile industry has faced a rapid growth not only in magnitude, but also in technology.

More than one billion vehicles are cruising around the world. The first hybrid vehicle appeared in 1997 when Toyota introduced Prius. Since that time, alternative power vehicles have been introduced commercially.

Technological innovation has upgraded the internal combustion engine quality and improved the electrical vehicles from depending on a small battery to being powered by an electrical motor.

The IEA is projecting more than 20 million electrical vehicles to be on the road by 2020. According to EU transportation policies, which enhance energy efficiency, a new passenger car had a weight-based corporate fleet average CO_2 emission regulation of 130 g/km by 2015.

The share of less than 130 g/km cars sold is already changing. For example, in Spain the share increased from 30–40 percent of vehicle sales in 2005 to above 50 percent in 2012.

EU and Japanese vehicle fuel economy standards are the most stringent at less than 6 1/100km, lowering to less than 5 a/100km by 2020. Globally, vehicle fuel economy standards are standardizing in line with the New European Driving Cycle test standards.

TABLE 3.3

Conventional versus Solar Power Plant

Power Plant Types	Pros	Cons
Conventional power plants	Economic	Emissions
	Reliability and availability	None- sustainability
Solar power plants	Clean (no emissions)	Additional cost for grid integration
	Sustainable	Additional cost for storage
	Maximize fuel exporting	High temp & dust versus efficiency
	Simplicity	Intermittency

On the electricity front, technology has played a magnificent role in shifting electricity generation from depending on fossil fuels to renewable energy. Table 3.3 presents the pros and cons for conventional power plants and solar power plants.

Solar and wind energies are the most feasible renewable energy sources. They satisfy the low-carbon requirements and reduce global warming effects. The key factors to be considered in adapting both types of renewable energy include sustainability, production cost, and operation and maintenance. However, to what extent could technological developments maintain efficient and effective electricity generation, distribution, and end-use consumption?

3.8 ECONOMICS OF ENERGY EFFICIENT TECHNOLOGIES

Energy efficiency and conservation have been at the heart of energy policy at national and international levels. Such policy with regard to energy efficiency and energy conservation has gained high priority, especially with the growing concern about climate change and energy security.

Measuring energy efficiency and energy intensity as an indication of energy conservation is the basic analysis used regarding the economics of energy efficient technologies.

Reaching acceptable levels of energy efficiency is a mandatory policy introduced by most governments. Technology should work hand-in-hand with policies and regulations to satisfy the required levels of energy conservation and efficiency; deploying technology and expanding the utilization of advanced energy systems in transportation, buildings, and electricity generation and consumption are essential to satisfy energy efficiency requirements and targets. This involves financial cost and its burden falls on industry and governments.

However, there is a wide spectrum of benefits from energy efficiency which include increasing economic efficiency through energy security and reducing energy-related public expenditure and improving the environment. From an international perspective, reduced greenhouse gas (GHG) emissions, natural resource management, moderating energy prices, and development goals are considered to be major deciding factors.

Lowering the energy demand can be discussed both at an international and national level, particularly for net energy-importing countries. Also, GHG emission reduction is crucial for national targets and climate change mitigation strategies. However, since international agreements about GHG emissions and their global implications became an important issue, they can be treated as being of international concern. GHG emissions became an important global concern after climate change in the last decade.

International organizations developed certain specifications and governments started to monitor their industries.

GHG emissions are reduced when energy efficiency improvements result in reduced demand for fossil fuel energy. Since energy efficiency measures are considered as the most cost-effective method, many climate change mitigation strategies are developed mainly with those measures in order to reduce GHG emissions.

If an aggregated international level is considered, less demand can reduce pressure on resources along with potential beneficial impact on prices for importing countries, in addition to overall resource management. For instance, energy efficiency can help to relieve pressure on a scarce resource in the context of peak oil and possible supply constraints. Demand reduction in several markets will decrease energy prices.

This situation may have positive implications on economic competitiveness for energy importing countries. Additionally, it would improve the affordability of energy services and the availability of resources. Improved energy efficiency is also crucial in terms of achieving economic and social goals in developing countries.

These goals involve improved access to energy services, exterminating poverty, improving environmental sustainability, and economic development. Advancing development in emerging countries in a sustainable way is a shared international goal with additional benefits.

The main cost elements that are involved in energy efficiency and conservation are the investment in technology development, which produce energy-saving systems and equipment. There is the direct cost of acquiring efficient lightning and advanced building systems.

This is along the with efforts in supporting technologies and policies for managing greenhouse gases and capturing CO_2, which will be lowered with more efficient use of energy.

The EU Energy Efficiency Directive requires EU states produce and pursue energy efficiency targets. These are central to implementing EU energy efficiency policy, which currently targets a 20 percent improvement in energy intensity by 2020.

Along with international efforts to develop a data and evaluation process for energy efficiency in public buildings and facilities, energy management information systems developed by UNDP are used as a tool for monitoring actual consumption of energy in public building. This system collects analysis and reports on data for energy consumption in public buildings and facilities in some selected countries.

The joint International Finance Corporation and initiatives have been established to commercialize energy efficiency finance. Climate investment funds have been one of the world's largest climate finance mechanisms since being established in 2008.

It collaborates with the Clean Technology Fund that focuses on financing energy efficiency projects and clean technologies' areas.

For example, there is a trading mechanism for delivering energy efficiency which covers policies such as White Certificate schemes, in place in several countries. There is a need for mechanisms to reflect the cost of pricing and energy price subsidies.

As a mandate, the United States in 2011 invested $7 billion in ratepayer-funded energy efficiency projects, producing an estimated 117 TWh of energy reduction.

In an attempt to increase the endorsement of higher efficiency appliances, the United States established the Environmental Protection Agency with specific programs to certify and endorse appliances and buildings. These procedures and practices have been disseminated among OECD countries.

PART IV Forces Shape Energy's Future

4.1 CHANGES AND CHALLENGES

Changes and challenges are both interrelated and interact at the same time. The question remains as to which causes the other. Changes might create new challenges which need to be taken into consideration. Most of the time, new challenges need new technical and administrative changes to be implemented with some modifications in the existing systems in order to achieve the necessary balance between changes and challenges.

Figure 4.1 depicts energy trends and the forces that shape the future prospects of energy – including oil as the main source of energy.

Within each force, there are changes and challenges, which flow into energy trends. For example, energy prices will interact with all other forces in a cyclical way.

Some of the changes are substantial and some are trivial, it depends to what extent they effect the current energy situation. Subsequently, this will have an impact on the type and importance of such challenges.

4.1.1 CHANGES

Since the earliest days of using energy, either in its primitive forms or in its conventional forms, there have been continuous changes in the processes of dealing with energy during production or utilization. However, it is hard to draw a line of distinction between a change and its challenge. Accommodating any change must involve challenges; therefore, a change is a challenge in itself. The most noticeable and influential change is the technological change.

Over history, technology development has affected most energy use. Technology, with is on-going change, brings with it different and difficult challenges which can be hard to cope with. Technological developments have several economic and social implications, which require introducing new polices and regulations as well as new technical and logistical support.

This is of course needs the application of new financial systems, and the restructure of politics and administration. Other important changes which are political may either change the system – as in the case of the break-up of the Soviet Union in 1989, resulting in the adoption of a capitalist system – or a free open economic system (market system).

The collapse of communism was a major change which brought with it the challenge to transfer from an economic system to another system with a different political

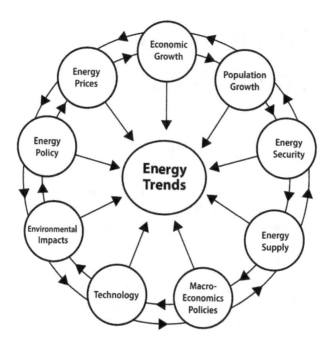

FIGURE 4.1 Changes and challenges.

platform and agenda. The most important change – and closely related to energy – is the market structural change.

This can be seen clearly in the oil business: once OPEC was founded, it diminished the monopoly power of the international oil companies. This created a new market structure with a challenge for oil producers and consumers to operate within a new market setting, which had implications on energy pricing and trading.

Aside from the changes of technology developments and market structural changes, there are major revolutionary developments in information technology (IT) as a result of implementing new innovations in electronics and communications.

This impacts all aspects of energy from supply to the end use. New information technology has changed the methods of energy production for all forms of energy. Along the way, it has changed energy distribution and marketing techniques. From the demand side, it has altered the ways energy is consumed, through new devices, from electric cars to coffee makers. The most important factor of change is seen in the new political and economic systems that emerged after the cold war era and the collapse of the former Soviet Union.

A new world order has evolved with a new feature, which is one economic and political power: the United States of America and the international relations system became "unipolar". These big changes have modified the future of energy, especially in the cases of oil and gas. The big effect is having a free market or open economy philosophy which dominates world economic behavior. It is noted that new economic powers have emerged with a strong influence on energy consumption and trade, such as China and India.

The result is more influence of market forces, and new countries and regions becoming oil producers and consumers. Production has increased with lower costs, and new oil trading companies appeared in the oil market such as Vitol, at the expense of OPEC national oil companies' market shares.

Thus, consumption in return has also increased, which has caused new patterns of petroleum production and consumption. OPEC should take the lead in oil trading which would increase its market share and strengthen its economic power. This will of course not be accomplished without full support from its members by taking a collective effort in this direction.

The biggest change is the emergence of unconventional oil and gas which has reduced oil and gas prices and moved attention from the traditional oil suppliers.

An immediate effect is an increase in US oil production which has resulted in lifting the export ban on US oil.

4.1.2 CHALLENGES

The biggest ongoing challenge is from technological developments. Technology has introduced new forms of energy and modified the systems and processes of energy production and consumption and changed the tastes and habits of people. Changing technology is a real challenge, which implies moving from one form of energy to another.

For example, moving from coal to oil was a big challenge for both industry and household energy consumers. It involved changing industrial and transportation systems as well as the ecological system.

Even the heating and coaling systems changed by using oil rather than coal. These challenges repeatedly happened in the case of moving from oil to natural gas, and from fossil fuels to renewable types of energy, such as wind and solar energy.

This challenge of shifting from one type of energy to another because of technological developments has reshaped the energy interface in general, as well as OPEC's future in particular. As a matter of fact, it has reshaped the world economy, and oil and energy markets.

Another important challenge is geopolitics. Developments in geopolitics exert pressure on energy markets and flow of energy, which affect the energy mix. It is a challenge to deal with new production and consumption patterns of energy. New production and consumption regions have emerged and most likely new trading and transportation systems will be developed. In addition, new forms of energy and new producers and consumers will appear accordingly.

This is apart from the impacts on energy prices. More recently, this can be seen in the case of oil with respect to the political problems in the Middle East.

An example is the invasion of Kuwait by the former Iraqi regime of Saddam Hussein in 1990 and the invasion of Iraq by the United States in 2003. In its aftermath, there were the Arab Spring events in several Middle Eastern Arab countries, which erupted in 2011. This geopolitical turmoil is an acute challenge for energy's future.

It has impacted energy prices, especially oil and gas. It also created situations of uncertainty in different energy markets. The world economic conditions are considered a real challenge that will shape the present and future of oil and energy.

Global financial crises are good examples. The most recent example is that of the 2007–2008 world financial crisis. Due to this, the world financial system was about to collapse followed by worldwide economic recession. Economic growth is a key factor in determining energy investment and consumption.

It is a challenge for both energy producers and consumers to have high economic growth to accelerate energy's demand and ultimately investment in expanding energy production. This challenge will enhance future energy security and involves new energy investment risks with new taxes and subsidies, and schemes.

The most recent challenge is the environmental impact of using fossil fuels – or as claimed – the production and consumption of conventional energy. The issue of protecting the environment gained a lot of momentum in the 1990s with great debates from all concerned parties, politicians, economists, energy experts, industrialists, and academics over three decades.

This international debate turned out to be inconclusive. An outcome of this debate is the issue of climate change, and global warming, which has received a great deal of attention lately. It has the potential to become a real challenge in determining energy's future, and more specifically oil's future, which will affect the future of OPEC.

4.2 FORCES AT WORK

The forces which shape the future of energy and OPEC range from changes in energy types and its prices (including oil) to technology developments. All forces form or lead to what are called "energy trends". These forces vary in their impacts on the future of energy, depending on its relevance and importance. The forces at work at any time of high importance are energy prices.

Energy prices are a very dynamic force which affect the shifting from one energy form to another. The price per unit of energy is the benchmark, which leads to the breakeven point or the breakeven price which determines the economic viability of any energy source. For example, moving from conventional oil to tight oil (shale oil) is based on a breakeven price of $70/bbl. Therefore, OPEC has increased its production at a lower world oil price, which results in a decline of shale oil production. However, because the oil price is so dynamic and technology was spreading, the production of shale oil was sustained by oil companies even at $40/bbl and did not decline except at the beginning of a price decrease. Without going into the economic technicality of measuring the breakeven point, still, the concept is used to analyze the decision to shift from one project to another or from fossil fuels to any type of renewable energy. Knowing energy renewable costs will give an idea of the breakeven price of each form of energy renewables. Several years ago, the wind cost in the US was $11c/kWh (US costs per kilowatt hour), and solar costs were $17c/kWh (Fattouh, et al., 2019). With technological development, these costs are assumed to be lower. This has of course become a real challenge for OPEC oil producers to meet the competition between oil and renewables.

Another important force that is in effect and influences energy trends is economic growth and conditions. This influential force is due to the interactions between fiscal and monetary policies and the integration between political and economic factors in the world at large or in a country. On the world scale, economic growth is highly essential in determining the supply and demand for energy. With an increase in economic growth, and other forces assumed to be almost constant, the energy supply increases in response to the rise in demand for energy. This applies for all kinds of energy and countries (Alsahlawi, 2013). The relationship between economic growth and energy use is important to determine energy intensity and therefore, the energy efficiency use-to-economic-growth ratio. There is a debate on what causes the effects of energy consumption on economic growth – or vice versa. It is found that an increase in energy demand usually increases economic growth in most developed OECD countries. In any case, energy intensity has certain policy implications on energy conservation and better use of energy. However, it is known that high energy intensity indicates costs to convert energy into economic growth.

It is very clear that changes in economic conditions can be attributed to international and regional financial situations as, for example, the global financial crisis of 2008. These crises have severe ramifications on world energy and energy markets. Economic conditions are impacted by geopolitical events in the world or in a certain region. The performance of economic growth globally or in a country has clear impact on energy demand and supply. For example, in the beginning of the 21st century, China witnessed rapid economic growth which was reflected by its demand for energy in general – and oil in particular. Energy policy is a direct force in mapping energy's future. This force is a result of dealing with two important forces: technology and the environment. Any energy policy that can be adapted and implemented with regard to technological and environmental regulations will change energy's future to the extent that it will alter the landscape for energy and the energy production and consumption pattern.

The most immediate force, which usually alters the production and use of energy, is the noticeable movement from conventional energy to renewable energy in a power generation. The energy mix in electricity's generation has changed from conventional types of energy such as oil fuels and coal to new types like liquefied natural gas (LNG) and renewables, or nuclear energy. This shift has been enhanced by the desire to save the environment through limiting carbon emissions and controlling the climate change phenomena. This process of energy substitution is affected by the changes in prices of different kinds of energy. However, the price trends and cost structures are very important determinants in deciding which fuel to use. For example, wind power is considered clean energy, and lately become competitive compared to other form of renewables, when it comes to cost and efficiency. A newly emerging force is that of digital technology. It is widely neglected in energy sectors from production to consumption and trading and operations. Digital technology is used in data collection, monitoring energy supply and demand, and controlling the operation for maintenance and other activities including trading. Last but not least, socioeconomic forces are changing future prospects for energy. Through demographic changes and population growth, the demand for energy is varying from region to region and from country to country, which affects demand for energy globally. Needless to say, when economics and political variables interact with demographic structures, the shape of energy's future is affected and causes change, as happened during the Arab Spring violence in the Middle East in 2011 and after.

PART V OPEC'S Existence

5.1 CHALLENGES TO OPEC

OPEC will face several challenges in the near and distant future. The most important challenges are keeping oil as the dominant source for energy. With all the challenges that face energy's future, preserving the advanced status of oil in the energy mix is a formidable challenge. The threat to shake oil's prime position is coming from renewable energy, especially with environmental protection pressure.

This is in addition to pressure on the use of oil from technological developments which make use of non-fossil fuels such as nuclear energy. For example, in transportation there is the trend now to move to electric vehicles.

There is the ongoing challenge of oil supply security, which should assure meeting future oil demand growth. Most studies forecast world oil demand to increase by 1.5–2.7 percent annually, given reasonable world economic growth of about 2 percent per year. In 2020, world oil demand is expected to reach 110 mbd.

Most of this demand – or about 35 percent of it – could be met by OPEC. This indicates that the market share of OPEC has been eaten up by non-OPEC producers; nevertheless, the OPEC oil reserve – from all its members – accounts for almost 70 percent of the world oil reserve.

This fact places a challenge on OPEC to supply more oil at a reasonably low price to compete with their non-OPEC rivals and with unconventional oil producers, as well as with non-oil sources of energy on the basis of low cost of oil production in most oil countries. Securing an oil supply to meet oil demand at a very competitive price is a real challenge for OPEC.

A closely related issue to the security of the oil supply is the need for investing financially in expanding the production capacities of OPEC member countries.

This spare capacity will guarantee the availability of oil at any time at an acceptable price to both producers and costumers.

This is the test that OPEC faces due to the challenging political and economic conditions of member countries or all over the globe.

The geopolitics of OPEC are pertaining to its member countries' geopolitics, which have been changing over the years based on economic, demographic, and political conditions. More recently, the Arab Spring events which erupted in 2011 in several countries in the Middle East and North Africa are examples. This region constitutes the major and most influential OPEC members, especially the Gulf countries, which make up most of the oil in the world with almost 60 percent of proven world oil reserves.

Those oil-producing countries were very vulnerable to the Arab Spring activities. Those sudden uprisings were ignited by social and economic problems with major political implications. Such events have destabilized the region and the world at large and affected oil price fluctuations with a lasting impact on OPEC. Many argue that oil prices' sharp and fast jumps over the past years are caused by political events rather than internal market forces.

Such events were the Arab–Israeli War in 1973, the Iranian Revolution in 1979, and subsequently the 1980–1988 Iraq–Iran War and the following various Gulf wars, until the US invasion of Iraq in 2003. This is, of course, including the political disputes with possible wars, such as the US boycott of Iranian oil as a result of the Iranian implementation of its nuclear program.

This might support the argument that oil price movements and frequent oil supply disruptions are caused by external factors rather than by internal oil market behavior.

However, political instability in oil regions is inspired by oil as a vital source of energy, which allows external powers of oil-importing countries to interfere in the politics and economics of the oil producers. Yet, this cannot be justified on any grounds, even on the basis of "peak oil" theory, which might explain oil scarcity or security of an oil supply principle.

5.2 OPEC'S STRATEGIES

Every five years, the OPEC Secretariat prepares a long-term strategy which provides a vision for the future of OPEC. The strategy centers around OPEC's objective to coordinate oil policies among its members' internal oil price and output distribution.

It takes into consideration the current and future challenges that face OPEC. The long-term OPEC strategy concentrates on three main issues: market stability, security of supply, and demand security. It is noted that this strategy has changed throughout OPEC's history.

From day one, OPEC's announced strategy was to have a fair price for its oil, which requires serious negotiations with major oil companies. The follow-up strategy at that time was to penetrate the oil industry in order to gain international recognition as an intergovernmental organization that deals with the economics of oil and presents the voice for developing countries.

Building such an image was at the heart of OPEC's strategy which was consistently achieved by capitalizing on its members' massive oil reserves and oil production capacities.

Over the years, OPEC gained some control over the production and pricing of its oil and tried to secure a fair income. Keeping this in mind, OPEC adopted the role of market regulator. This strategy has been implemented during periods of excess oil supply by applying flexible pricing systems. This has put high pressure on OPEC to be responsible for stabilizing the oil market, and ultimately the world economy.

OPEC aims to build a strategy grounded on three pillars: price stability, security of oil supply, and ensuring the importance of OPEC as a market stabilizer and regulator. The three pillars are interrelated and their implementation is attempted simultaneously. The future OPEC strategy should concentrate on the existence of OPEC in an ever changing world, where the energy landscape has been changing dramatically.

Oil share in the energy mix is shrinking with the potential increase in a renewable energy share. Therefore, OPEC should design a new strategy of supplying oil with special specifications in terms of price competitiveness and desirable environmental specifications. This will require new policies with regard to pricing strategy and new approaches toward dealing with renewable energy types and climate change as an ongoing battle.

OPEC's future strategy should emphasize OPEC solidarity and modify its regulations to accept new members. Reviewing and updating OPEC rules has to be an essential and continuous function of the OPEC strategy to achieve higher efficiency, better management, and fast responsiveness to any urgent political or economic development. The persistent OPEC strategy should reinforce the role of oil as a dominant, sustainable, and secure fuel.

Strategically speaking, OPEC has to struggle for its existence and penetrate the energy business as a key player. This cannot be attained without adhering to its main objectives and the full cooperation and coordination of its member countries. Over the course of OPEC's history, it has succeeded in achieving many important objectives – against all odds.

The future strategy of OPEC should be built on the basis of having OPEC as the leading player in oil business with a clear vision of its future investments on expanding production capacities and having green energy.

5.3   OPEC'S FUTURE

Based on the current and future changes and challenges that face OPEC, its future may be anticipated. Each of the strategic challenges has multifaceted impacts on OPEC's existence. The short and long-term strategies that were built internally by OPEC are aimed at justifying its importance and existence as a market stabilizer. However, it could be questioned as to just how long and how much power it will have to perform this role in energy markets and the world economy.

The future of OPEC is certainly determined by changes in the structure of the world oil market. Before OPEC, the oil market was controlled by international oil companies, which were dominating the oil industry, including OPEC's oil production and distribution.

The oil price, however, was low due to the prevailing oil surplus and price cut by oil companies and the former Soviet Union, in order to encourage European countries to buy oil. The evolution of OPEC somehow created competition in the world oil market with more influence from spot prices, and with spot futures and options emerging.

There is more coordination between oil companies and OPEC members on pricing matters. This has emphasized the sovereignty of OPEC producers and control over their oil with improving oil prices and lessening the influence of oil companies; new oil regions were discovered, and new oil producers emerged, with some eventually becoming members of OPEC.

Furthermore, OPEC countries are able to go into downstream integration in refining marketing and distribution activities.

This means more power for OPEC in terms of having a higher market share with respect to oil production and to a certain degree to the refining sector, especially for the major OPEC members such as Saudi Arabia. However, the challenge for OPEC is to go for refining most of its produced crude oil. This is a major step to be taken in the future for economic diversification purposes and increasing the value added of crude oil.

Of course, this requires massive investment, and upgraded process technology, as well as the need for marketing skills and better locational advantages. Table 5.1 presents refining capacities in OPEC at Total and other leading oil producers.

It also presents the refining capacities within OPEC countries. As noted, some countries built their refineries at home and some bought a share of existing facilities in different consuming countries, such as in the case of Saudi Arabia with Texaco. (Alsahlawi, 1990). Another important factor in determining OPEC future is the oil demand and supply structures. Regarding the demand, the price movements and levels of world economic activity are the most important elements that shape future oil demand and OPEC's future.

Reviewing the demand structure with respect to oil price, we can see that during the period before OPEC's existence, the price was low and demand, therefore, was high. During the high prices of the 1970s, demand was growing but at a low rate. Demand increased with the declining oil prices of the 1980s and 1990s.

However, the lower oil demand because of the rise in oil prices caused a shift somehow from oil to other fuels such as gas and to other alternative energies, not to speak of energy conservation, which even rationed energy consumption and reduced demand further. This is one of the current challenges that OPEC is facing in the future, which will be the management of the demand for oil and the increase of substitution by other efficient fuels.

As far as the impact of economic growth on oil demand and effect on OPEC's future, it has less immediate impact as compared to oil price's rapid effect. Over the periods of high economic growth, it was found that demand for oil increased.

TABLE 5.1
Refining Capacities in OPEC and Leading Countries, 2010–2020, (mbd)

Year Country	2010	2018	2020
United States	17.6	18.8	20.3
China	12.4	14.9	16.1
Russia	5.6	6.7	7.5
OPEC	9.6	12.8	13.2
Saudi Arabia	2.3	3.5	4.4
Iran	1.3	2.1	2.5
Venezuela	1.8	2.0	2.2

Source: Author's estimates based on various sources.

The trend of increasing demand has been the norm whenever oil prices are low and the world economy is expanding.

However, during periods of recession, as experienced in some years of the 1990s and since 2008 – and most recently with the case of coronavirus – the demand for oil would almost be stagnant with a dark future for OPEC.

From the supply side, non-OPEC output has emerged as a key element in determining the future of OPEC. The price increase of the 1970s led to an increase in non-OPEC production at the expense of the OPEC market share.

However, recent analysis argues that the non-OPEC oil supply is about to be depleted and has peaked (Alsahlawi, 2010). On the other hand, it is seen that there is more potential for oil discoveries in OPEC countries, as well as in non-traditional areas such as Venezuela and Brazil. This apparently contradicts the oil peak theory (Hubbert, 1971), which predicts oil production first increases, reaches a peak, and then declines. In general, oil supply, whether from OPEC or non-OPEC, is always affected by price movements and technological developments. Based on such factors, Table 5.2 shows the distribution of the world oil supply and reserves by regions as well as the reserves-to-production ratios (R/P).

It is noted that most of the oil supply comes from the Middle East region, which is the core of OPEC. Non-OPEC production, on the other hand, has increased since the 1980s onward and most analysts thought that its peak would be reached in the second decade of the 21st century (Franssen, 2005). However, the prediction of peaking oil has been a matter of speculation, even with more advanced technologies and better geological knowledge. The future of OPEC will not be discussed without exploring the prospects of the world oil supply which clearly depend on world oil reserves. For more than 50 years, world oil reserves were increasing annually, by almost 3 percent at normal consumption rates. This is given the fact that OPEC controls more than 70 percent of the world's proven oil reserves over the same period.

As far as the remaining life of world oil reserves is concerned, the reserves-to-production (R/P) ratio as an easy measure indicates 45 years. This is as shown in Table 5.2: the R/P ratio points out that the world oil reserves will run out in almost 45 years at current production rates and state of technology.

Forecasting the future oil supply will depend greatly on oil price trends and, to a certain degree, the proven oil reserve. Nevertheless, high oil prices will encourage higher oil production even in high-cost areas. Under conservative price scenarios and content reserves with reasonable changes in the state of technology and no major changes in political or legal systems, the world oil production will increase by less than three percent annually.

The outlook for non-OPEC production substantially affects the future of OPEC; however, oil production from OPEC and non-OPEC depends, of course, on the oil price profile as well as on technological developments which will most likely enhance world oil production.

Another factor that certainly affects the future of OPEC is geopolitical change. This is despite the fact that OPEC member countries have tried not to let their political differences affect their decisions and common objectives on the basis that OPEC

TABLE 5.2
World Oil Supply (mbd), Reserves (bnb), and Reserves-to-Production Ratio (R/P) by Regions

Year	1970			1990			2020*		
Region	P	R	R/P	P	R	R/P	P	R	R/P*
North America	10.9	49.8	12.9	8.6	31.8	10.1	9.0	32.0	12.5
Latin America	5.2	26.2	13.8	6.9	122.8	48.6	10.2	13.8	43.5
Europe	7.7	68.2	30.2	14.5	75.0	13.5	18.5	14.0	20.0
Middle East	13.8	336.2	66.5	16.1	662.5	112	23.5	750.0	85.5
Africa	6.1	51.2	31.4	6.1	58.6	26.7	11.2	120	33
Asia and Pacific	1.5	17.2	21.4	6.3	34.2	14.8	7.5	42.1	15.2
Total World	45.2	548.8	33.0	59.2	985.0	45.4	78.0	1225	45.0

Source: OPEC Annual Statistical Bulletin (several issues)

* Author's estimates

is an economic organization. However, political relations within OPEC and with other countries are important in shaping the future of OPEC and the world oil market.

5.4 ENERGY'S FUTURE

OPEC's future cannot be separated from the future of energy in general. Energy mix in the future gives a clear comparison between different kinds of energy including oil in terms of production and consumption. In the future, oil will constitute more than 30 percent of total energy produced with almost the same percentage for consumption over the next 30 years. This means that oil will remain a major source of energy in the future with clear implications for OPEC's existence.

So, all in all, it depends on oil demanded versus oil supplied. Oil price trajectory and variation in world economic growth will determine oil supply and demand. The future prospects of other forms of energy as oil substitutes depend on many factors. The determinant factor is the market oil price where the demand for oil and its supply intersect. Other factors that affect the development of different energy types are technology, capital cost, and political motives. Figure 5.1 presents forces that shape the future of energy.

These factors are interrelated with each other and affect each other. For example, capital cost influences the decision process of developing energy and the technology which is used in energy development and its production as well as consumption.

Obviously, energy's future is determined by the need for energy diversification given the economic, political, and technological conditions. Yet, the price of oil is a key factor in making the decision to move away from oil as principal fuel to another form of energy as renewable energy.

However, moving away from oil to other forms of energy doesn't mean eliminating the need for oil but it might reduce slightly the share of oil in the energy mix and it might increase the energy efficiency in general.

The question that should be asked is whether OPEC will exist in the future, even with the continuous need for oil. The real challenge with the future of energy is how reliable and sustainable the different kinds of energy will be.

Other important elements are the affordability of these types of energy and their safety from an environmental point of view. The future of energy should fulfil at least the needs of the two main sectors: power generation and transportation.

There is no doubt that both sectors have already developed from technical and financial perspectives and it is expected that new developments will be implemented in the future, such as the wide production and use of electric cars for transportation and smart grids in the area of electricity generation and transmission.

The energy sources with the most potential for the future are solar and wind due to the important fact that both are abundant with positive environmental impacts.

The key setback is the problem with storing the energy efficiently and effectively. However, the main challenge concerning energy's future is the sustainability to ensure the supply will meet the demand.

Another challenge is the ability to cope with technological innovation in renewable and non-renewable energies. The environmental challenge in mitigating greenhouse gas emissions and controlling climate change is a very pressing issue in determining the future of energy. Last but not least, the changing geopolitics surrounding energy is a reminder of the fact that it will shape its future.

5.5 OPEC'S GLOOMY FUTURE

Despite the fact that OPEC owns around 80 percent of the world crude oil reserve, and controls about 40 percent of the world oil production, OPEC seems to be losing its power. However, with all the geopolitical crises that have faced OPEC since its foundation including the Arab Spring events of 2011 and their consequences, OPEC has survived its role as an oil market stabilizer.

Nevertheless, the oil market has changed structurally with new oil producers entering the market and with new trading practices such as the emergence of future markets and speculations, yet OPEC remains more or less influential and is capable of settling its members' differences and has been able to accommodate its internal rivalries over the years. However, it is noted that OPEC has become more responsive to market forces rather than dictating its decisions on the market.

The perception that OPEC is a cartel or monopoly which controls the market has disappeared and is seen as merely an illusion. In reality, OPEC lacked the ability and the means to enforce production quotas among its members and failed to set an example for discipline internally – but this shows OPEC's diversity.

It has become clear that OPEC's future is highly affected by world and regional geopolitics, but OPEC has survived past tensions from within and outside the organization.

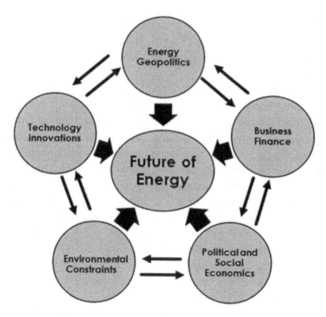

FIGURE 5.1 Forces shape the future of energy.

However, there is a great doubt about OPEC's ability for maneuvering politically and economically in such a dynamic new world.

This of course might create some ambiguity and a hazy future for OPEC. Another element that would add to the gloomy picture of OPEC's future is the trend for privatization in general and for the privatization of OPEC national oil companies (NOCS) in particular.

Privatization has been a way for achieving better managerial and economic efficiency and reducing budget deficits for governments. Privatization has become a global phenomenon since the 1990s; however, OPEC countries have been latecomers. Privatization of NOCS is a very sensitive issue not only for OPEC countries but for most of the world, which has to do with sovereignty and control problems.

However, due to the changing oil market and world political structures as well as the need to finance governments, few OPEC members are considering selling their NOCS publicly as IPOs. However, OPEC should take the initiative to trade in oil, but it remains to be seen how it will be accepted by the members.

Saudi Arabia, Kuwait, Nigeria, and Venezuela are examples. The most recent privatization, Saudi Aramco's five percent IPO was proposed during the period from 2016 to 2017, then in 2018 (Katona, 2017).

On November 17, 2019, Saudi Aramco priced its IPO at $8.5 per share which meant that Aramco would have a market valuation of $1.7 trillion as the most valuable listed company. In addition to global economic conditions, oil market and price volatility, restructuring of international oil industry, and geopolitical tensions are the most important determinants of OPEC's future. Based on past behavior and a business-as-usual scenario, OPEC has proven to be useful in regulating the market and stabilizing oil prices.

However, current political and economic changes have major implications on all markets, not only the oil market. This can lead to restructuring the oil industry toward more company privatizations and less dependence on OPEC. The more privatization of national oil companies, especially in OPEC countries, the less control it owns over petroleum resources and less the influence of OPEC.

Therefore, it is to be expected that OPEC'S role is less important and the organization might be even abolished. This conclusion will be aggravated and enhanced by any geopolitical tension, especially in OPEC regions or by members – all of which possibly causes oil disruptions that cannot be replaced by OPEC. This indicates less coherence within OPEC.

With the new market developments and less control of oil producers over their resources, and more possible political or military disputes among OPEC member countries or internationally, there will be more tendency for OPEC fragmentism.

Furthermore, OPEC no longer acts as a cover for its members to pass on its policies or desires against superpowers or most influential oil importers or exporters, such as the United States or Russia. On the other hand, OPEC is no longer used as a "shoulder to cry on", or someone to convey the superpowers' wishes or policies. These new trends are due to the high level of transparency in the oil market and international politics which allow powerful countries to dictate their polices regardless of OPEC – all of which makes OPEC less effective and redundant.

5.6 A WORLD WITHOUT OPEC

OPEC was established in 1960 to break up the cartel of the major international oil companies that dominated since the beginning of the twentieth century. Since then, OPEC has been justified – on the basis of preserving the member countries' rights – to decide their oil production and obtain a reasonable oil price.

Over the sixty years of OPEC, it has struggled to achieve objectives: there were many years where OPEC lost its ground and showed weakness in controlling the oil market.

OPEC also witnessed over the years clear interference in its business by external powerful countries such as the United States, Russia, and those of Europe. If we assume that OPEC will be abolished in the coming years, then what might the world look like?

Without OPEC the world will no longer have OPEC to blame for its economic problems or its failures to balance the oil market when the market is in turmoil. Without the existence of OPEC, it is expected that bilateral relations among countries will prevail. Whatever the outcome, it is noted that even with OPEC, there has been no solidarity between members and no adherence among them as to the decided production quotas and prices.

This will continue when OPEC is absent from the scene and it becomes preferable for each member country to act independently based on its own interests. Such independent acts would be suitable for any member of OPEC, especially for major producers such as Saudi Arabia – or even for a big non-OPEC producer, such as Russia or the United States.

The consequence of such free production: prices will be driven downward to competitive levels – or even lower. A worse scenario would be having a price war among oil producers until both small and large producers go out of business.

This situation happened in OPEC during the 1980s, and oil prices were maintained, not because of OPEC but due to Saudi Arabia, when it acted as a swing producer by reducing its production.

However, restoring low oil prices to an acceptable level in the absence of OPEC would be the responsibility of each producer as is the usual case with the presence of OPEC. Therefore, it might be argued that OPEC to a certain degree is immaterial in regulating the oil market, and the world can live without OPEC. The key question is: what would the market structure be without OPEC?

There are three possible outcomes: first, a tendency to have perfect competition, where each producer is a price taker with no power to control the price, given that the world oil supply equals the world oil demand at a price level that reflects the marginal cost of production.

The second outcome is to have one or a few giant producers without any formal cohesion among them deciding a price and acting as a price leader, while the rest of the producers follow and compete over the market shares.

The third outcome is the case where there is a dominant big producer with low costs of production or large scale of economies who sets the price and its production level, which hence maximizes its profits. The other oil producers will take the price as given and distribute the remaining output from the total industry output between them, based on their cost structures.

In a world without OPEC, market forces will dominate, especially with existing several international oil companies, and newly emerging sizeable national oil companies (NOCs).

Both kinds of companies are now competing internationally in different ways: the size, the degree of integration, and the geographical market distribution. This indicates that the market is the regulator – rather than OPEC.

PART VI Side Issues

6.1 QATAR AND NATURAL GAS

In the early 1970s, Qatar was exploring for natural gas, even though it was a small OPEC member with less than half a million barrels per day of oil production at that time. This move attracted attention from OPEC and the world media. The media tried to be a leading observant in covering this major energy development. The story started when a large deposit of natural gas was found in the north east of Qatari territorial water in 1971, which was given the name "North Field" (Figure 6.1). The updated estimated reserve is around 800 trillion cubic feet (cu ft) which makes Qatar the world's third largest natural gas reserve after Russia and Iran. In 1987, the decision was made to bring the North Field into production. The current production rate is about 6.0 billion cu ft per day.

It is mentioned in the above figure the notion of an "Arabian Gulf" which is used as a common term in the Arab world, but which runs against the official Iranian name of "Persian Gulf". This led OPEC to reach an agreement or a common understanding to use the neutral term, "the Gulf".

Qatar now is considered as a mega world gas supplier, mainly to the Far East region. In the process of developing the North Field, several major multinational companies in construction and industry were involved, transforming it into an all-round state-of-the-art project. The idea was to have Qatar as a key natural gas exporter and a local and regional gas supplier to different key industries such as petrochemical and desalination plants – not only in Qatar but also in other Gulf Cooperation Council (GCC) countries. In the process of developing the Qatari natural gas industry, there have been many by-products including human resources development and technology transfer given the most sophisticated facilities during production and distribution, as well as the exporting of terminals. In 2018, Qatar was the leading world exporter of liquified natural gas (LNG) with about 105 billion cubic meter (bcm) and the second worldwide natural gas exporter. This gives Qatar more power in the Gulf as a source of natural gas and within OPEC as a bargaining card in oil production decisions. This, however, encouraged Qatar to leave OPEC in January 2019 (it had joined in 1961). The decision to leave OPEC has some political roots, when Qatar relations with Saudi Arabia and UAE are considered.

6.2 OIL SPILL IN THE GULF

One of OPEC's focuses is on environmental issues. In 1991 the world and OPEC devoted their attention to an environmental problem which happened along the Saudi East Coast of the Gulf as a result of a massive oil spill. Figure 6.2 presents the location and the path of the spill.

The environmental impact speaks for itself, while the economic impacts are obvious, as the path of the oil spill reached the water intakes of the desalination

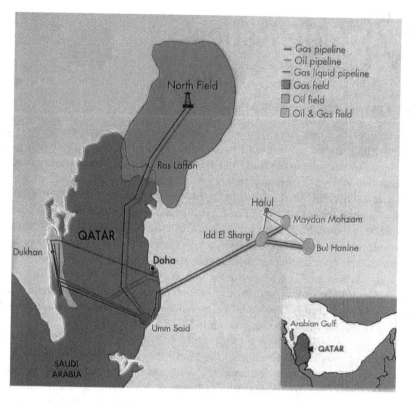

FIGURE 6.1   The North Field Development. Source: OPEC Bulletin, September 1992.

plants that supply sweet water to the capital of Saudi Arabia, Riyadh. In addition, the spill damaged the water cooling facilities of the petrochemical industrial complex in Jubail.

Because of the size of the problem, and Saudi Arabia being an OPEC member, OPEC devoted the section of a country sketchbook in its monthly OPEC Bulletin of October 1991 to this problem. In order to deal with this catastrophic problem, Saudi Arabia formed a team of experts from different concerned government organizations supported by international agencies and independent experts.

The first task of the rescue team dealing with oil spill was to collect and gather the oil to move it toward the shoreline in an effort to trap or pump it out. After oil collection, the next process was skimming to remove oil from the surface of seawater. Utilizing tidal movements and the wind direction were also used for collecting oil.

These significant operations saved the marine life as well as rare birds, as shown in Figure 6.3 below.

6.3   KUWAIT OIL FIRES

After the Iraqi invasion of Kuwait on August 2, 1990, and the failure of diplomatic efforts, Saddam Hussein refused to pull out of Kuwait. The Allied Forces led by

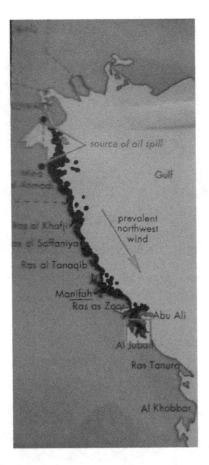

FIGURE 6.2   The source and path of the oil spill. Source: OPEC Bulletin, October 1991.

the United States declared war against Iraq, on January 17, 1991 under the name of "Desert Storm". The retreating Iraqi troops set fire to over half of the Kuwaiti oil wells in addition to other oil facilities, see Figure 6.4. The immediate damage released a huge amount of oil to the Gulf, with black smoke covering the sky of Kuwait and neighboring countries. It was a real environmental disaster that resulted in air, water, and ecological pollution. Its effects extended to the whole region. To extinguish the burning wells, a collective effort was organized with the idea that the fires would take two years to put out. Many international fire extinguishing companies were hired from all over the world. Several methods were implemented and used to put out the fires.

The most common method was to spray sea water with high power hoses at the fire base. The first well fires were put out in April 1991 while the last well fires were extinguished on November 6, 1991. On this occasion, Kuwait celebrated the event and invited high officials and key representatives from all countries and international organizations such as OPEC. Because of the sensitivity of the issue, since Iraq and

FIGURE 6.3 Oil drenched bird found on the shore. Source: OPEC Bulletin, October 1991.

FIGURE 6.4 Oil well fires in Kuwait. Source: Getty Images, 1991.

Kuwait are members – and were involved – OPEC was reluctant to attend. This was to that extent that the Secretary General welcomed the invitation and forwarded OPEC's congratulations, but he preferred not to attend himself.

The author, on his capacity as OPEC Director of Information and Public Relations, attended the celebration on behalf of the Secretary General. He commented on Kuwaiti media when recognizing the first-class achievement by Kuwait in extinguishing the fires. Despite the magnitude of the problem of the Iraqi invasion of Kuwait with social, political, economic, and environmental consequences, OPEC took a neutral position in the whole matter because of both countries, Iraq and Kuwait, being members with equal weight. OPEC stressed the notion that it is an economic organization and it does not have anything to do with politics or military actions.

The root of the problem started when both countries disputed over the Rumaila Oil Field and Iraq claimed that Kuwait was slant drilling in that field. What mattered to OPEC was that Kuwait was exceeding its production quota at that time, which annoyed other members in general and Iraq in particular – which aggravated the problem between Iraq and Kuwait. Whatever the justification for the Iraqi invasion of Kuwait, it should not have been to destroy international protocol by occupying an independent country.

6.4 OIL PRICE CRASHES

The financial setback because of the coronavirus and its negative effects on economic activities attributed to the sharp decline in oil demand and in oil prices. This is the most recent crash in oil prices. However, the previous crash was in the early 1980s when the oil supply was increased by OPEC and non-OPEC producers, which caused a drop in the oil price to less than $10 per barrel in 1986, and a world oil glut. See Figure 6.5.

This crisis was not a financial crisis per se but a weak oil market caused by financial disorder. To rectify the oil market, Saudi Arabia took the initiative to restrain OPEC output by reducing its production on the expenses of its market share and played the "swing producer" role until its output level reached about 3 mbd in order

FIGURE 6.5   Oil price crash of 1986 (OPEC is drowning in sea of oil). Source: Greg Stoicoiu.

to safeguard the oil price and hold it up. This strategy proved to be wrong: the Saudi market share was eaten up and failed to increase the price.

On the contrary, it encouraged other oil producers to increase production because of the improved price. As a matter of fact, what brought the market to stability was the increase in OPEC production – including Saudi production – in a way to eliminate producers of high-cost oil from continuing production. Luckily, oil demand improved which sustained oil prices.

The second oil price crash happened in 1997, when again, Saudi Arabia wrongly adapted the strategy of increasing oil production – on this occasion during the Asian financial crises. The crises began due to high foreign debt and unstable currencies in the Southeast Asian countries. The crisis was aggravated by floating currencies and the currencies' devaluation. With appropriate policies orchestrated by the International Monetary Fund (IMF) and World Bank (WB) based on a case-by-case approach, the crisis was over in 1999. The immediate effect of the crisis was on the demand for oil that hurt Gulf oil producers such as Saudi Arabia, Kuwait, and the UAE who are key OPEC members. With low demand and a surplus in the oil supply, the prices plummeted to less than $10 per barrel.

OPEC again – and mainly Saudi Arabia – failed to deal with the problem as a result of lack of cooperation among OPEC members, and also due to miscalculation of the size of the financial problem.

The fourth oil price crash was during the worldwide economic and financial crisis of 2007–2008, which was severe and beyond control. In terms of oil market, the crisis was an exogenous factor caused by credit risk and the collapse of financial institutions. It started in the US by depreciation in the "subprime home mortgage". As a result, stock markets dropped worldwide, many businesses went bankrupt, and private wealth vanished, and eventually led to a great recession which spread all over the world, noticeably in the US.

As far as OPEC is concerned, oil prices overreacted in the beginning of the crisis due to wide speculation in the oil market by oil traders. The OPEC reference price reached as high as 140/b in July 2008 and then dropped to around 30/b by the end of the year. This drop in oil prices was caused by the low world economic growth (recession) which lowered oil demand. Nevertheless, this wave of low prices continued until the crash of late 2014.

The combination of the geopolitical and the financial risks and the poor market management by OPEC caused this trend of low oil prices to continue until recently, when Saudi Arabia and Russia managed to coordinate their output to restrict OPEC and non-OPEC supply. Unfortunately, with the coronavirus effect on global demand, prices crashed again and became worse when Saudi Arabia and Russia entered into a price war. Such a price war is not justifiable economically or politically.

6.5   OPEC AND IRAN–IRAQ WAR

The Iran–Iraq war started in late September 1980 when Iraq invaded Iran on the basis of border disputes between the two countries. The war ended in August 1988 without significant gains for each country. Both Iran and Iraq are OPEC founders and major oil producers and exporters. However, OPEC acted as an economic organization without intervening in politics and the warfare between Iran and Iraq. The only thing affected was OPEC production during the war. Once the war stated, oil production fell to around 3mbd and continued to drop by almost 7 mbd from its level of 31 mbd in mid-1979. The production loss was substituted mainly by Saudi Arabia and other OPEC members. On the other hand, oil prices jumped over the period from August 1979 to August 1981 to over $30 per barrel or $100 per barrel in real terms for 2019 prices. This was the second price shock because of the Iranian Revolution and the war between Iraq and Iran, after the first price shock of 1973. The reported reason for the war was border differences, but the real reason was to control the spread of the Iranian Revolution to other Gulf countries, as it was announced by the Iranian leaders that they wished to export the revolution to the Gulf as well as to the rest of the world including, of course, their neighbor Iraq. The price of the war was very expensive for both countries and other Gulf countries – not only the loss of human life but also in material loss and damage in addition to the financial expense.

The negative impacts from social and geopolitical points of view were massive in the region, which has been reflected in the stability of the Middle East. As far as the oil market is concerned, the oil disruption from Iran and Iraq created a state of uncertainty with regard to oil prices and oil supply security. OPEC was confronted with several factors that affect the oil market. There was poor planning and a lack of control of the oil supply, especially in the light of increasing the supply from non-OPEC countries. Even with the fears of the war, oil prices dropped to their lowest levels in 1986. The OPEC market share was eroded, almost halting the oil supply from Iran and Iraq and reducing the oil supply, as OPEC was playing swing producer to sustain oil prices.

The superpowers, the United States and the Soviet Union (USSR), held different positions in the war. The United States supported Iraq in order to counter the threat of Iran on its behalf and undertook its mission to protect its allies in the Gulf. On the contrary, the USSR played the game neutrally, but provided Iraq with military assistance in terms of weapon and intelligence. This position continued until the war ended in a way, the US maintaining Iraq as an ally and limiting Iranian involvement in the Islamic states of the USSR, especially Afghanistan which was already under its invasion. When it came to oil and OPEC, the administrative requirements within OPEC were distracted even in the issue of appointing a Secretary General. The awkward situation between Iran and Iraq affected the internal relations among the members and delayed many of the important decisions including production and price agreements. OPEC succeeded in maintaining oil supply security for the world, because there were excess production capacities within key members such as Saudi Arabia to compensate the lost oil from Iran and Iraq. However, there were discounted oil sales from both countries to individual countries, oil companies or traders, which found its way to the market and reduced the general price of oil – this is of course, aside from oil smuggling and illegal deals for military services and weapons. Export outlets in North Iraq such as the Iraq–Turkey pipeline and the north of the Arabian

Gulf were always under major attack from Iraq. At the same time, Iran was suffering from Iraqi attacks on its oil export and refining facilities in the south of Iran such as Kharg Island, the main oil export terminal. Most of the oil facilities in the Gulf, including the Strait of Hormuz, are vulnerable to disruption by sabotage or by air and naval attacks. Generally speaking, all petroleum plants and facilities in the Gulf states are under risk of attack, which puts pressure on the oil market and oil prices. Figure 6.6 shows the petroleum facilities in the Gulf and how these facilities were vulnerable during the Iran–Iraq war.

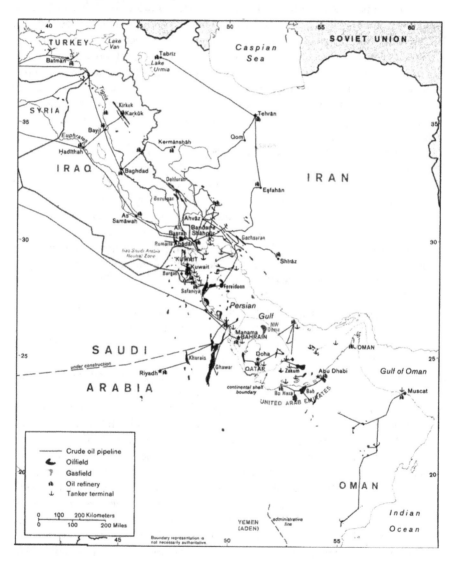

FIGURE 6.6 Petroleum industry facilities in the Gulf. Source: CIA-RDP84B00049R001403400038-4 Approved for Release 2007/04/24.

6.6 CORONAVIRUS (COVID-19)

On December 31, 2019 the World Health Organization (WHO) announced that Wuhan, a city in China, was attacked by a new virus. That was the first city in the world infected with the virus, which was given the name coronavirus (COVID-19). The disease became a global pandemic with over half a million confirmed cases in the last week of March 2020. The US, Italy, and Spain had the most infected cases outside of China.

In the UK, there were 14,579 confirmed cases and 758 deaths as of March 27 (Reynolds, March 2020). Most countries around the world were put under lockdown and shut down their social activities. People who had the symptoms of "fever and dry cough" were quarantined for 14 days.

The world seems unlike that before the coronavirus, and cities all over the world look empty and resemble cities of ghosts rather than those of normal people. Homes turned into life graves like the Egyptian pyramids: full of everything, on the assumption that Pharaohs will return back to life after death. "Just stay home and be safe" became the theme in all countries. The norms of life have changed, and the world economy has been affected negatively as well as domestic economies going into a deep recession.

It is a state of war albeit without using physical weapons for destruction, but the coronavirus causes damage to health as well. After seeing the economic and social damage, some countries lifted the lockdown, but the second wave of coronavirus forced some countries to return to lockdown.

The coronavirus has affected human relations as well as international relations. What is happening in dealing with the coronavirus is a social, political, and economic coup without military forces. It is expected there will be the implementation of new rules and regulations at an international relations level, which might change intergovernmental policies. It has been a change to a new world order, with immediate economic, social, and political ramifications such as an expected long-lasting global recession with immediate collapse of stock markets and a dramatic fall in oil prices.

This is what really concerns OPEC, due to the stagnation of the world economy and the sluggish oil demand, which are attributed to the coronavirus, that has stopped usual economic activities and caused a huge layoff of employees. This, of course, has increased unemployment rates worldwide and in many major countries, such as the United States. OPEC was not be able to do anything with regard to controlling the coronavirus except for the individual efforts of each member country to prevent the disease from spreading over its territory. Figure 6.7 envisages how OPEC saves the world; however, the stagnation of economies and lockdown of economic activities including transportation and air travel have reduced fuel demand and put pressure on OPEC to deal with the excess oil supply and to manage the oil market.

OPEC tried, on the other hand, to manage the oil supply by limiting the output, relying on "OPEC+" cooperation in order to sustain oil prices. In this regard, OPEC put high hopes on Saudi Arabia and Russia, the largest world oil producers (other than the US) to cooperate and try to curb oil production. Both countries are fighting for their market shares, in spite of low oil prices, to the degree of entering into a price war.

FIGURE 6.7   OPEC and Covid-19. Source: Greg Stoicoiu.

Recently, continuous talks between the two countries including the United States to reach a compromise in order to reduce "OPEC+" production substantially at least by more than 10 mbd in a hope to support oil prices in order to reach almost $40/b. From an oil perspective, usually wars increase oil prices in a fear of losing oil supply resulting in high demand but with the coronavirus, oil prices dropped as oil demand declined because of the world economy operating at a low level. This explains the difference between war economics and pandemic economics.

Wars stimulate demand for certain products and with an oil supply shortage, oil prices usually rise. However, with less demand and excess supplies, oil prices will definitely go down. There is the second wave of the coronavirus which necessitates the unity of OPEC+, led by Saudi Arabia, to stabilize world oil prices. More recently, most countries have lifted lockdown policies, but without a pledge to provide a cure. Yet, several countries are now racing to produce vaccines but still there are some debates about its effectiveness and safety.

There is a general feeling that the coronavirus will stay for the coming year without a definite answer on its impact on general health and world economies – as well as on the social and political activities which altered all existing systems, be they social, economic, or educational.

PART VII Wrap-Up

7.1 WHAT IS THE POINT OF OPEC?

OPEC is a unique international origination: if it were not there, it would have been invented. The oil industry was controlled by the major international oil companies (the Seven Sisters). The oil market was monopolized in setting the oil price, and decisions regarding the output were taken by those companies with full coordination among them, along with well-integrated operations from the exploration for oil till the last stage of marketing and distribution, along the way passing through the production and refining. Because of the new geopolitics after World War II, international oil companies changed their priorities and strategies. This led to new economic and industrial relations between the companies, which created mistrust among them and a lack of market control. As a result, prices went down to unprecedented levels to less than $2/barrel in nominal terms.

In such a foggy environment of the oil industry, the oil producers in the Middle East and Latin America took the initiative to raise the issue of unfair low oil prices and subsequently low economic returns for them as host countries. After implicit and explicit discussions and negotiations among them – and between them and oil companies – they decided to establish a league of oil producers in order to coordinate a common position regarding oil production and oil price in the face of the major international oil companies' position. That league was subsequently called OPEC.

In September 1960, the founders Venezuela, Saudi Arabia, Iran, Iraq, and Kuwait announced the birth of OPEC; international oil companies and western countries met OPEC with a cool reception, nonchalance, and high skepticism. Now OPEC celebrates its 60-year anniversary and is still in existence, but its effectiveness and influence remain debatable. During the first decade of OPEC's existence, most of its activities were diplomatic efforts to build its image as an intergovernmental organization dealing with oil economic issues and representing a few oil producers from developing countries.

In the following decades OPEC strived in securing the interests of its members and their ownership over its petroleum resources.

OPEC's role in the oil market depends on market conditions, i.e. whether the market is characterized by oversupply or high demand. OPEC is supposed to deal with each situation as a market regulator. However, in most cases OPEC failed in playing this role given the divergence among its members' interests, and conflicting external factors.

Certain issues such as the environmental and the geopolitical imposed itself on OPEC. These issues put high pressures on OPEC and in most cases, OPEC failed to cope.

The high prices of the 1970s gave incentive to explore for oil in non-traditional areas known for high cost such as the North Sea and Alaska's North Slope. The new discoveries flooded the market with oil which reduced oil prices to low levels of less than $10/b in 1986. This was the beginning of eroding OPEC's market share for the benefit of non-OPEC new producers. The emergence of new oil suppliers put OPEC's influence and role into question. However, OPEC tried to take the lead in managing the world oil supply and stabilize the market through more coordination with non-OPEC producers. Usually, in a weak market, OPEC takes the burden by reducing its output and losing its market share, while in a high price market, everybody enjoys high production and high prices. Based on this scenario, many economists and analysts started to ask: what is the point of OPEC? Or, what would the future of OPEC be?

These questions are becoming more apparent with the development of unconventional oil and renewable energy sources. The big question – what is the point of OPEC? – can be answered by referring back to the objectives and accomplishments of OPEC. As pointed out, the main objective of OPEC is to coordinate oil production policies among its members. On the other hand, OPEC accomplishments or achievements could be summarized as one thing: establishing a club for net oil exporters from less developed countries. It is clear that OPEC alone cannot manage or control the oil market – there is always the need to bring non-OPEC oil producers to the table of negotiations which always fail and end up by OPEC taking the burden and compromising on its production. Over the years, OPEC became a media platform for its oil and energy ministers as, for example, the Saudi oil minister Ahmed Zaki Yamani during the 1970s, who became a world celebrity. Sometimes, OPEC was used as a political arena for its members to show off or as a political champion for settling political disputes among some members or among members and other countries from outside OPEC or between OPEC and other organizations such as IEA. OPEC, as a typical international organization, sometimes appears boring and bureaucratic without eliminating corruption but sometimes appears interesting and stimulating.

As far as OPEC's future is concerned, it is considered by the oil market as a safety valve: when the oil market becomes saturated with oil, OPEC is approached or will take the initial to withdraw some of its oil to balance the market and restore oil prices to acceptable levels. Another future benefit for OPEC is to have it as a source of information on oil industries – not for OPEC members only but for the whole world. Another benefit is to consider OPEC as an umbrella where its members can negotiate their positions regarding economics and other issues, such as climate change or energy technologies.

In summary, OPEC in the beginning was trying to find its way and protect its existence among oil market players, mainly the major oil companies. Yet, OPEC in the 1970s, was taken advantage of by its non-Arab members who were producing oil in compensation of oil cut by the Arab members for imposing the oil embargo of 1973 against the United States and any other country who was supporting Israel. Figure 7.1 shows the Arab oil embargo of 1973.

In the 1980s, new oil producers started to pump their oil into the market, then, OPEC's market share was eaten up by non-OPEC producers, while OPEC and leading producers, such as Saudi Arabia, reduced their output in order to protect OPEC's market share. But unfortunately they failed to reach that objective.

FIGURE 7.1   Arab oil embargo of 1973. Source: Greg Stoicoiu.

During the following decades – and until now – OPEC almost lost hope as a market regulator and became a losing horse. Therefore, one might argue that what is the benefit of OPEC, and why not have a world without OPEC? On its 60th anniversary on September 14, 2020, OPEC faces a critical situation of crashing demand of oil and prices because of the coronavirus, i.e. the COVID-19 pandemic. This collapse of oil prices has shed light upon the future of OPEC and its ability to control the oil market – or whether to survive the ever-changing world economy and oil market.

7.2   ENERGY'S FUTURE

Energy's future determines the future world economy and its social and political developments. Historically speaking, energy has played a major role in shaping the world economy and human civilization. Since the last century, oil has been the most important type of energy which has enhanced world economic development and been the origin of certain political motives, to the extent of causing military conflicts among nations. However, oil price fluctuations and technological development allowed for the finding of energy alternatives other than oil. Yet, energy's future does not stop at discovering new forms of energy but also developing new methods and techniques in production and consumption of energy, including, of course, oil and gas.

The integration between energy and technology has shaped the future of energy. This has been attained through a systematic way of converging the technology mix into an energy mix. In this matrix, the new revolution of information technology (IT) acted as a catalytic converter which has changed current energy applications and contributed to the future of energy systems. With recent innovations, IT has altered energy production, distribution, and consumption methods. Changing technology is a real challenge in determining future energy. The impact of digital technologies on energy is at hand. Digital energy affects energy law, energy demand and supply. As a result of the coronavirus and protection measures, the world is moving to digital operations in all sectors including energy systems of productivity, security, sustainability, and risk management.

This implies how technology development is effectively moving away from one form of energy to another and to use energy more efficiently and successfully. The future of energy has no limit: renewable energy has high perspectives. It is a matter of how much it costs and how reliable and sustainable it is. It is also a question of when societies and systems are ready to make the transfer from fossil fuels to renewables or from one type of renewable to another. This might take time, but all depends on how much it costs. Now oil and gas consumption – and even coal – have been enhanced by applying new technologies of exploring and production and gasification, in the case of coal. The advancement of oil and gas exploration and development technologies such as fracking in oil production have increased US oil production to unexpectedly high levels.

In the near future, fossil fuels will remain the dominant energy, but with no exception, all countries will change to renewable energy for the obvious reasons of having a cleaner, more efficient, and sustainable energy. Recently, some developing countries like India have been moving to rely on electricity to be used in vital sectors such as transportation. The new trend is indicating that power generation in the future will rely more on alternative renewable energies such as solar, biofuels, wind, hydro, and hydrogen, and to a lesser degree on nuclear energy.

In summary, the world's future energy will be mainly determined by the cost factor where the bottom line is how much it costs to produce and consume energy. Furthermore, investment will be needed for technology development – even at the stages of research and development. As a matter of fact, the cost will affect the manufacturing and producing of new energies, be they conventional or renewables. Calculating the cost should include the ramifications and impacts of the environment, political and social economic factors on the whole world, or on an individual country. Nevertheless, assuring the sustainability of the new technologies will be added to the costs. However, guaranteeing the readiness of the social and political systems to adapt for future energy is vitally important when considering the whole transfer costs, as based on the principle of "economies of scale".

Bibliography

Adelman, M. A. *The World Petroleum Market*. Johns Hopkins University Press, 1972.

Ahrari, Mohammed E. "OPEC: The Failing Giant". 1986. https://books.google.com.sa/books? id=f5sfBgAAQBAJ&pg=PA78&lpg=PA78&dq=opec+threatens+world+economy& source=bl&ots=NvRrCSualk&sig=ACfU3U2P6L5ddAwHsdQXnoXgwwqjEUhV 5w&hl=en&sa=X&ved=2ahUKEwiUysjtuNXkAhVE5uAKHSl8CTw4PBDoATAFe gQICBAB#v=onepage&q=opec%20threatens%20world%20economy&f=false.

Alsahlawi, M. A. "Oil Price Changes and Non-OPEC Oil Supply: An Empirical Analysis". *OPEC Energy Review*. Vol. 131, pp 11–20, March 1989

Alsahlawi, M. A. "OPEC – Aims, Achievements, and Future Challenges". *The Courier, No. 130*. 1991.

Alsahlawi, M. A. "Qatar Prepares to Move Into the Gas Export Market". *OPEC Bulletin*, Vol. XXIII, No. 8. September 1992, OPEC, Vienna.

Alsahlawi, M. "The Real Prospect of Non-OPEC Oil Supply," *Journal of Energy And Development*, Vol. 18, No. 2, PP. 175–180, Spring, 1993.

Alsahlawi, M. A. "The Future Role of Oil in the Global Energy *Mix*". *OPEC Review*, Autumn, pp. 297–308. 1994.

Alsahlawi, M. A. "Oil and the U.S. Dollar, Structural and Empirical Analysis". *The Journal of Perspectives in Energy*, Vol. III, No. 4. . pp. 435–440, 1994–1995.

Alsahlawi, M. A. and Elbek, M. "An Alternative Oil Pricing Currency to Improve OPEC's Balance of Trade". *Journal of Energy and Development, Boulder, USA*, Vol. 22, No. 2, pp. 187–197. 1997.

Alsahlawi, M. "The Dynamics of Oil Inventories," *Journal of Energy Policy*, Vol. 26, No. 6, pp. 461–463, 1998, UK.

Alsahlawi, M. A. "An Alternative Oil-pricing Currency and OPEC`s Foreign Assets". *Journal of Energy and Development*, Vol. XXXlll, No.1., pp.. 81–90, 2009.

Alsahlawi, M. A. "Oil Price and the U.S Dollar: A Survey of The Empirical Relationship Estimates and Alternative oil-Pricing Currencies". *Journal of Energy and Development*, Vol. 36, No. 1–2 PP. 45–62. (Autumn, 2010 and Spring, 2011).

Alsahlawi, M. A. "The Future Prospect of World Oil Supply: Depletion of Resources or Price Trends". *OPEC Energy Review*, June. 2010.

Alsahlawi, M. A. "Measuring Energy Efficiency in GCC Countries Using Data Envelopment Analysis". *The Journal of Business Inquiry*, Vol. 12, pp. 15–30. 2013.

Alsahlawi, M. A. *Structure of the Oil and Gas Industry in 'Petroleum Economics and Engineering'*, Eds H. K.Abdel-Aal and M. A.Alsahlawi, GRC Press, New York. 2014.

Beck, Martin. "OPEC+ and Beyond: How and Why Oil Prices Are High". *E-International Relations*, January. 2019. https://www.e-ir.info/2019/01/24/opec-and-beyond-how-and -why-oil-prices-are-high/.2019.

Benjamin, Zycher. *OPEC: The Concise Encyclopedia of Economics*. Library of Economics and Liberty. 2019. Accessed September 16, 2019. https://www.econlib.org/library/ Enc1/OPEC.html.

Bern, Gianna. "OPEC's Role During the Financial Crisis 2008 to 2009 - Investing in Energy: A Primer on the Economics of the Energy Industry [Book]". 2009. https://www.oreilly.com/ library/view/investing-in-energy/9781118128381/OEBPS/c12-sec1-0005.htmL.

Brew, Gregory. "The United States, OPEC, and International Oil". *Oxford Research Encyclopedia of American History*, May. 2019. https://doi.org/10.1093/acrefore/9780199329175.013.719.

Bye, B. et al. *Energy Technology and Energy Economics: Analysis UF Energy Efficiency Policy in Two Different Model Traditions*. CREE, OSLO. 2018.

Carnelos, M. "OPEC`s Slow Decline". *EXPERT Opinions*. 2019. http://valadaiclub-com/a/highlights.

Daniel, Yergin. "The Prize: The Epic Quest for Oil, Money & Power". 2008. https://www.amazon.com/Prize-Epic-Quest-Money-Power/dp/1439110123.

Defterios, John. Hawks vs. doves in Vienna, pp. 1–3 June 9, 2011, http://edition.cnn.com/2011/BUSINESS/06/08/defterios.opec.oil/index.html

Directorate of Intelligence. "OPEC and the USSR: The Oil Connection". *An Intelligence Assessment, Sanitized Copy Approved for Release 2011/05/19*. 1983.

Dorraj, Manochehr. "*Will OPEC Survive?*". *Arab Studies Quarterly*, Vol. 15, No. 4, pp. 19–32. 1993.

Econlib. "Who Caused the August 1990 Spike in Oil Prices?". June 30, 2014. https://www.econlib.org/archives/2014/06/who_caused_the.html.

ESCWA. *The Impact of the Global Financial Crisis on The World Oil Market and Its Implications for The GCC Countries*. Economic and Social Commission for Western Asia (Escwa). 2009. http://www.regionalcommissions.org/crisis/escwacri1.pdf.

Fattouh, B., R. Poudineh, and R. West. "The Rise of Renewables and Energy Transition: What Adaptation Strategy Exists for Oil Companies and Oil-Exporting Countries?". *Energy Transitions*, Vol. 3, No. 1–2, pp. 45–58. 2019.

Fisher-Vandlen, K. et al. "Technology Development and Energy Productivity in China, Energy Economics". Vol. 28, No 5–6, pp. 690–705. 2016. https://doi.org/10.1016/J.eneco.2006.05.006.

Franssen, H. I. *The Future of Oil = Will Demand Meet Supply? Demand Implication of Peak Oil and the Geopolitics of Middle East Oil*. American Meteorological Society Energy Briefing, Washington, DC. 2005.

Gately, Dermot et al. "Lessons from the 1986 Oil Price Collapse. Pdf". 1986. Accessed September 16, 2019. https://www.brookings.edu/wp-content/uploads/1986/06/1986b_bpea_gately_adelman_griffin.pdf.

Gately, Dermot, M. A. Adelman, and James M. Griffin. "*Lessons from the 1986 Oil Price Collapse*". *Brookings Papers on Economic Activity*, Vol. 2, p. 237. 1986. https://doi.org/10.2307/2534475.

Gillingham, K., R. Nowell, and K. Palmer. *Energy Efficiency Economics and Policy*. NBER Working Paper, No. 150-31. 2009.

Granger, Morgan M., and D. W. Keith. "Improving the Way, we Think About Projecting Future Energy Use and Emissions of Carbon Dioxide". *Climate Change*, Vol. 90, pp. 189–215. 2008.

Gross, Samantha. "The 1967 War and the 'Oil Weapon'". June 5, 2017. https://www.brookings.edu/blog/markaz/2017/06/05/the-1967-war-and-the-oil-weapon/.

Hanewald, Christopher. "The Death of OPEC? The Displacement of Saudi Arabia as the World's Swing Producer and the Futility of an Output Freeze". *Indiana Journal of Global Legal Studies*, Vol. 24, No. 1, pp. 277–308. 2017. https://doi.org/10.2979/indjglolegstu.24.1.0277.

Herring, H. "Energy Efficiency: Analytical View". *Energy*, Vol. 31, No. 1, pp. 10–20. 2006.

https://inflationdata.com/articles/inflation-adjusted-prices/historical-crude-oil-prices-table/.

https://www.facebook.com/meo.news/node/669706.

https://www.fool.com/investing/2017/03/29/-why-isnt-russia-a-part-of-opec=aspx.

https://www.forbes.com/sites/stratfor/2013/12/05/the-future-of-opec/.

https://www.forbes.com/sites/stratfor/2013/12/05/the-future-of-opec/#376c546312dc.

https://www.theatlantic.com/magazine/archive/1983/03/the-cartel-that-never-was/306495.

Hubbert, M. K. *The Energy Resources of the Earth, Energy and Power.* Freeman and Co., San Francisco. 1971.

IEA. *Key World Energy Statistics*, Analysis and Data statistics, Annual edition of data services, 2014

Immanuel, Wallerstein. *Saudi-Iranian Collaboration: A Forgotten Story" International Energy Agency (IEA), 2014. 'World Energy Outlook'.* IEA. 2016.

Irawan, P. *'Understanding the Global Rivalry Between OPEC and IEA' A Note.* Vienna, Austria. 2012. https://www.facebook.com/meo.News/node/669706.

Jilani, Humza. "OPEC Close to Agreement to Open the Oil Taps – Foreign Policy". June 21, 2018. https://foreignpolicy.com/2018/06/21/opec-close-to-agreement-to-open-the-oil -taps-saudi-iran-russia/.

Kaltschmitt, M., W. Streicher, and A. Wiese. *Renewable Energy: Technology Economics and Environment..*, Springer, 2007.

Katona, V. "How to Successfully Privatize A National Oil Company". *Oil Price.com.* 2017. https://oilprice.com/Energy/Crude-Oil/How-To-Successfully-Privatize-A-National-Oil-Company.html.

Kawata, Y. and Kazuo, F. *Some Predictions of Possible Unconventional Hydrocarbons 'Availability Unit 2100'.* Society of Petroleum Engineers, Asia Pacific Oil and Gas Conference, 17–19 April, Jakarta, Indonesia. 2001.

Law, B. E., and J. B. Curtis. "Introduction to Unconventional Petroleum System". *AAPG Bulletin*, Vol. 6, p. 11. 2002.

Learsy, Raymond J. *Over a Barrel: Over a Barrel: Breaking Oil's Grip on Our Future.* Encounter Books, New York. 2007.

Lesourd, J. "Solar Photovoltaic System: The Economics of Renewable Energy Resource". *Environmental of Modeling & Software*, Vol. 16, No. 2. Pp. 147–156. 2001.

Mikdashi, Zuhayr. "The OPEC Process". *Daedalus*, Vol. 104, No. 4, pp. 203–215. 1975.

Minczeski, Sarah McFarlane and Pat. "OPEC vs. Shale: The Battle for Oil Price Supremacy". *Wall Street Journal*, April 18, 2019, sec. Markets. https://www.wsj.com/articles/opec-vs-shale-the-battle-for-oil-price-supremacy-11555588826.

MintPress, Newsdesk. "The Internal Politics of OPEC Threatens Global Energy Security". 2019. Accessed September 16, 2019. https://www.mintpressnews.com/opecs-internal-politics-threaten-global-energy-security/208090/.

Mitchel, J. "Oil Production Outside OPEC and the Former Soviet Union: A Model Applied to the US and UK". *The Energy Journal*, Vol. lS. 1994.

Naghdi, Yazdan, Soheila Kaghazian, and Nashibeh Kakoei. "Global Financial Crisis and Inflation: Evidence from OPEC". *ResearchGate*, January. 2012. https://www.researchgate.net/public ation/268353461_Global_Financial_Crisis_and_Inflation_Evidence_From_OPEC.

Offshore Technology. "From OPEC to Iran: Moments Which Have Rocked the Oil Industry". *Journal of Offshore Technology | Oil and Gas News and Market Analysis (blog)*, July 16, 2018. https://www.offshore-technology.com/features/opec-iran-moments-rocked -oil-industry/

OPEC. "OPEC: Brief History". 2019. Accessed September 16, 2019. https://www.opec.org/ opec_web/en/about_us/24.htm.

Opinion - Chinadaily.Com.Cn. "Is Saudi Arabia Planning to Leave OPEC?". n.d. Accessed September 16, 2019. http://europe.chinadaily.com.cn/a/201811/27/WS5bfc82e5a310ef f30328b295.html.

Pelletiere, Stephen C. "Iraq and the International Oil System: Why America Went to War in the Gulf - Google Books". 2019. Accessed September 16, 2019. https://books.google .com.sa/books?id=EQ1cpB65KR4C&pg=PA151&lpg=PA151&dq=opec+doves+and

+hawks&source=bl&ots=XYCm3unzoi&sig=ACfU3U2OH0mc-K4N8cN__25qOJ7t
HeC_0w&hl=en&sa=X&ved=2ahUKEwjXrNTctdXkAhWIohQKHb4qDTs4ChDoA
TAFegQICBAB#v=onepage&q&f=false.

Pierru, Axel et al. "OPEC's Impact on Oil Price Volatility The Role of.Pdf". 2018. Accessed
September 16, 2019. https://www.researchgate.net/profile/Tamim_Zamrik/publication/
323411328_OPEC%27s_Impact_on_Oil_Price_Volatility_The_Role_of_Spare_Capacity/
links/5b3506cb4585150d23dd8d48/OPECs-Impact-on-Oil-Price-Volatility-The-Role-
of-Spare-Capacity.pdf.

Pierru, Axel, James L. Smith, and Tamim Zamrik. "OPEC's Impact on Oil Price Volatility:
The Role of Spare Capacity". *The Energy Journal*, Vol. 39, No. 2. 2018. https://doi.org/
10.5547/01956574.39.2.apie.

PoPP, DC. "The Effect of New Technology on Energy Consumption". *Resource and Energy
Economics*, Vol. 23, No. 3, pp. 215–239. 2001.

Princeton Energy Advisors. "OPEC Oil Production During the Iran-Iraq War". 2019.
Accessed September 16, 2019. http://www.prienga.com/blog/2018/12/13/opec-oil-
production-during-the-iran-iraq-war.

Research Division. "The Impact of Current and Proposed Environment Measure on OPEC
Oil". *OPEC Secretariat*, Vienn. 1992.

Reynolds, M. "What is Coronavirus and What Happens now. It is a Pandemic?". *WIRED*, 27
March, 2020.

Roz, Liston. "OPEC Oil Producers Trying to Offset Iran-Iraq War Losses - UPI Archives".
2019. Accessed September 16, 2019. https://www.upi.com/Archives/1980/10/02/OPEC-
oil-producers-trying-to-offset-Iran-Iraq-war-losses/7935339307200/.

Schiller, M. "The Economics of Wind Energy". *Petroleum Accounting and Financial
Management Journal*, Vol.29, No.3, pp.55–81, 2010

Scott, K. "Wind Power Planning: Assessing Long-Term Costs and Benefits". *Energy Policy*,
Vol. 33. No.13, pp, 1661–1675, 2005.

Seymour, I. *OPEC Instrument of Change*. OPEC. Palgrave Macmillan, 1980.

Sieminski, A. "The $200 Billion Annual Value of OPEC`s Spare Capacity to the Global
Economy". Commentary, KAPSAKC. 2018.

Simkins, B., and R. Simkins. *Energy Finance and Economic: Analysis and Valuation, Risk
Management, and the Future of Energy.*, The Robert W. Kolb series in Finance, 15
Pages2013.

Sorrells, S. *The Economics of Energy Efficiency: Barriers to Cost-Effective Investment*.
Edward Elgar Publishing, 2004.

Soytas, U., and R. Sari. "Energy Consumption, Economic Growth and Carbon Emissions:
Challenges Faced by an EU Candidates Member". *Iconological Economics*, Vol. 68,
No. 6, pp. 1667–1675. 2009.

Spero, Joan Edelman Edelman, and Jeffrey A. Hart. *The Politics of International Economic
Relations*. Cengage Learning, 7th edition, Boston, US, 2009.

Tagliabue, John. "Stubborn Strategist: Sheik Ahmed Zaki Yamani; Squeezing Opec - And
the U.S. - The New York Times". *New York Times*, April 1986. https://www.nytimes.
com/1986/04/13/business/stubborn-strategist-sheik-ahmed-zaki-yamani-squeezing-
opec-and-the-us.html.

Tahmassebi, Hossein. "The Impact of the Iran-Iraq War on the World Oil Market". *Energy,
OPEC and ASIA: The Changing Structure of the Oil Market*, Vol. 11, No. 4, pp. 409–
411. 1986. https://doi.org/10.1016/0360-5442(86)90127-1.

Tyler, Patrick. "Yamani Replaced by Saudis - The Washington Post". *Washington Post*,
October 1986. https://www.washingtonpost.com/archive/politics/1986/10/30/yamani-
replaced-by-saudis/a2eec13e-91fa-4e17-aa88-b8cccb3361a3/.

Wall Street Journal. "Saudi Arabia, OPEC's Anchor, Ponders a Future Without the Cartel - WSJ". 2018. Accessed September 16, 2019. https://www.wsj.com/articles/saudi-arabia-opecs-anchor-ponders-a-future-without-the-cartel-1541703893.

Wilson, Rodney. "Review of *Review of The Myth of the OPEC Cartel: The Role of Saudi Arabia; OPEC and Future Energy Markets, OPEC, Instrument of Change*, by Ali D. Johany, Opec, and Ian Seymour". *The Economic Journal*, Vol. 91, No. 363, pp. 767– 771. 1981. https://doi.org/10.2307/2232849.

WIPO. http://www.wipo.int>academy. 2013.

World Bank. "Anatomy of the Last Four Oil Prices Crashes". *Commodity Markets Outlook*, April 2015 http://pubdocs.worldbank.org/en/40441444853733469/CMO-April-2015-Feature-Oil-Price-Crash.pdf.

World Intellectual Property Organization (WIPO). *Statistics on Patents*.CH-1211, Geneva20, Switzeland 2013.

Yergin,D. The Prize: *The Epic Quest for Oil, Money and Power*. Simon & Schuster. New York, 1990.

Zakariah, Muhamad Hasrul Bin. "The Oil Embargo Following the Arab-Israel War of October 1973: British Economic Experience and Reaction". *Journal of Middle Eastern and Islamic Studies (in Asia)*, Vol. 5, No. 2, pp. 92–120. 2011. https://doi.org/10.1080/19370679.2011.12023181.

Zakariah, Muhamad Hasrul Bin. "The Oil Embargo Following the Arab-Israel War of October 1973: British Economic Experience and Reaction". 2011. Accessed September 16, 2019. https://www.tandfonline.com/doi/pdf/10.1080/19370679.2011.12023181.

Zycher, Benjamin. "OPEC, by Benjamin Zycher". In *The Concise Encyclopedia of Economics*. Library of Economics and Liberty. 2019. Accessed September 16, 2019. https://www.econlib.org/library/Enc1/OPEC.html.

Index

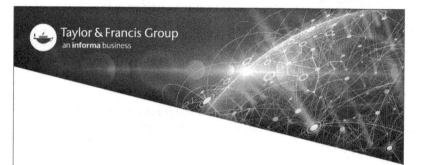